The Marketer's Guide to Successful Package Design

The Marketer's Guide to Successful Package Design

Herbert M. Meyers   Murray J. Lubliner

NTC Business Books

NTC/Contemporary Publishing Group

Library of Congress Cataloging-in-Publication Data

Meyers, Herbert M.
    The marketer's guide to successful package design / Herbert M.
    Meyers, Murray J. Lubliner.
        p.   cm.
    Includes bibliographical references (p.) and index.
    ISBN 0-8442-3438-9
    1. Packaging.   2. Marketing.   3. Packaging—Design.   I. Lubliner,
Murray J. II. Title. III. Title: American Marketing Association
HF5770.M46   1998
658.8'23—dc21   97-50260
                                                      CIP

Interior design and production by City Desktop Productions, Inc.

Published by NTC Business Books
An imprint of NTC/Contemporary Publishing Group, Inc.
4255 West Touhy Avenue, Lincolnwood (Chicago), Illinois 60646-1975 U.S.A.
Copyright © 1998 by NTC/Contemporary Publishing Group, Inc.
Printed in the United States of America
International Standard Book Number: 0-8442-3438-9
15 14 13 12 11 10 9 8 7 6 5 4 3 2

To our many colleagues among both marketers and consultants with whom we have had the pleasure and privilege of working. Enjoy!

Contents

Acknowledgments

A few words of thanks

This book would not have been possible without the generous contributions of several friends and associates. In this way, they have, in fact, become active participants in authoring this book, and we appreciate the encouragement that is implied in their offering their generous attention and help. They are

- Elliot Young, president of Perception Research Services, Inc., and a longtime friend, who reviewed Chapter 8, "Consumer Research: Navigating the Category," and whose considerable professional experience in packaging research assures that this chapter is on target

- Barbara Bowlend, another longtime professional associate and personal friend, whose valuable suggestions for Chapter 10, "Package Design for Special Markets," are based on her extensive activities as a designer of delightful packaging of products for children

- Eric F. Greenberg, specialist on packaging law, and Eric Wachspress, both attorneys at Chicago-based Bullwinkel Partners, Ltd., whose counsel and suggestions regarding Chapter 12, "Packaging and the Law," supplemented our own experience with legal matters

- Marc Gobé, president of Desgrippes Gobé & Associates, who contributed several photographs showing designs created by his firm

- Keith Davis, staff photographer at Gerstman+Meyers, who is responsible for many of the photographs in this book

- Molly Cuttingham, whose design skills helped develop and digitize the charts that appear in this book

- Ric Hirst, a living lexicon of structural design and technical know-how, who reviewed those pages in this book that describe the development of three-dimensional packaging, particularly Chapter 7, "Creative Development: Where the Rubber Meets the Road"

- Juan Concepcion and Sal D'Orio, who also reviewed Chapter 7 to confirm the procedures that describe graphic design development and finalization and the development of package design control manuals

- Bill Melin, computer guru at Gerstman+Meyers, whose digital expertise rescued our manuscript from numerous potential fiascos

- Maria Tricolla, administrative assistant at Gerstman+Meyers, who generously found the time in her hectic schedule to do a major portion of the typing of the manuscript for this book and put up with our many modifications and corrections

- Brian Falck of Heinz USA and Sid Sheggeby of Clorox, who, among others, encouraged us to go forward with this book by verifying its need

Our thanks go also to the many other professionals, acknowledged elsewhere in this book, whose wisdom we borrowed throughout the volume.

And, last but certainly not least, we wish to express our appreciation for the patience exhibited by our loving spouses, who frequently had to share our attention with our writing of this book. We know that ultimately they will participate in the pride we take in presenting *The Marketer's Guide to Successful Package Design* to our readers.

Introduction

This book is for you. Is there anyone reading these words who does not interact with packaging on a professional or personal level? Whether you are a brand manager striving to propel your products to a higher share of market, a marketing director who is setting strategies, an ad agency creative who is helping to mold a brand's image, or a designer who has been called in to "package" the product, you face the challenge of convincing the consumer that your product is better than the competitor's—that it is *the* best.

More than thirty thousand different products line the shelves, freezers, and aisles of our supermarkets and other outlets such as department stores, mass merchandisers, and specialty stores. And the proliferation grows daily. About two thousand new products are introduced each year, at a cost estimated to exceed one billion dollars. The consumer is bewildered by a continuous flood of new products, look-alike products, line extensions, and established products ensconced in unfamiliar packaging and sometimes wearing different names and identities. If an alien were to land in a typical retail mass-merchandise setting, he or she could conclude reasonably that the purpose of this jumble of packaging shapes and graphics is to deliberately confuse the consumer.

But that is precisely what a good package seeks to *avoid*—consumer confusion. The right kind of package links the consumer's mental process to an image already created by advertising, a personal experience, or a friend's referral and then triggers a buy decision. For those products that are not advertised and thus rely entirely on the package for creating a product image, the challenge is even more compelling. The right kind of package dispels uncertainty; it informs and educates; it appeals to the heart and mind of the prospective buyer, who makes a choice in less than ten seconds. The right kind of package says "Take me home; I'm worth the money."

But what is the "right kind of package"? How do you determine the physical configuration of the container? The materials? The shape? The graphic expression of the brand name? The visual format, the colors, and the styling of the text? The product's description? The supporting copy and selling message? The relationship of the parent company to the brand name? How do you select a designer to create the package? And what type of consumer research is most appropriate to determining whether the design is the right package?

When these aspects and the myriad of structural, visual, and verbal elements come together, they create the *image of the product* within the package. This is one of the most critical lessons about package design: The consumer buys the *product*. The package is the vehicle that helps create the *product image*. The consumer does not have the time or interest to analyze the package; he or she makes a decision on either a conscious or subconscious level about the efficacy of the product via the information that the package communicates. This is why the character of the product often guides the packaging solution.

More money is spent on packaging each year than on advertising. If we add the income of the manufacturers of the raw materials that go into a package (paper, metal, plastic, glass), the converters of the package itself (cartons, bottles, pouches, blister packs), the producers of the closures (caps, spouts, lids), the designers and researchers (independent consultants and in-house), the printers who bring the designers' work to life, and all others who originate and implement the process, the numbers exceed the gross national product of most of the world's countries.

Yet, too often packaging decisions are made in a crisis atmosphere—under pressure and late in the product marketing process. In this book we set forth and recommend procedures for treating package design as a rational business practice that merits the same logic and careful attention that strategic marketing and communications issues receive from senior management and from marketing and brand managers.

Costs of marketing continue to increase as a percentage of sales revenue across the entire range of packaged goods. This puts great pressure on everyone associated with conceiving and selling a product. For consumer-oriented products, packaging can make the difference between winning and losing the battle in the retail arena. The right package at the right time for the right product will boost sales and profits and establish a franchise in the marketplace; the wrong package can cause even a good product to fail. And the rate of failure, especially for new products, can devastate the bottom line as well as the careers of corporate managers and their design and advertising consultants.

There is no mystique to packaging; it is part of doing business, part of marketing strategy. On the pages that follow, we cut through the difficulties of creating "the right package" and provide the guidance that, as a professional, you need to develop the most effective package efficiently and economically—packaging that will powerfully support your product in the marketplace. Our goal is simple: to help you do a better job and *enjoy* doing it.

The Marketer's Guide to Successful Package Design

1 The Package Is the Product

Successful marketers understand that for the consumer *the package is the product.* The consumer sees and responds to the shape of the package, the recognition of the brand, the color and the words, and the graphic style and format and instinctively conjures up an image of the product. For the marketer, the key and the challenge are to turn that package into a powerful selling tool that achieves a competitive advantage. This is why the investments in packaging continue to increase each year and why package planning receives top priority among more and more companies.

Packages have many functions that contribute to their effectiveness as selling tools. In addition to *protecting* the product and making it easier to ship, distribute, and display at the point of sale, package design must serve a number of functions that are crucial to the *marketing* of the product by

- making sure that your product stands out among competing products
- providing brand and product identification that will be recognized by the consumer and will encourage trial and repeat purchases
- presenting your product in an appealing and attractive manner

- identifying product attributes and important information about the product
- identifying varieties, flavors, sizes, and other product specific factors
- enhancing your product by making the package easy to open, close, dispense, carry, and store

Achieving all or most of these objectives will maximize the probability of the consumer's selecting your product instead of a competing brand.

How can package design achieve a positive perception of your product? There are techniques by which package design can tell the consumer what's in the package and encourage a favorable judgment about the product. Let's examine some of the ways by which this can be achieved.

What Packaging Structure Tells the Consumer About Your Product

While most marketers are cognizant of the importance of graphics in their quest to build brand recognition and achieve brand equity, they may be overlooking one of the most fertile areas of building positive product imagery: the packaging *structure*.

The primary reason for the packaging structure is, of course, its advantages for both the consumer and the retailer in a wide area of functional features, ranging from product protection during warehousing, storage, and handling to store display, in-home use, consumer conveniences, and many more. Most important, you can outflank competitors through unique packaging structures and functional benefits that over the years will build a solid foundation of *equity* for your brand.

Webster's dictionary defines *structure* as "the interrelation of all parts as dominated by the general character of the whole." It is interesting that Webster's has a vision that goes below the surface by pinpointing elements of communication that may escape most of us. Webster's dictionary does not define *structure* in terms of technical components, as most of us would most likely do, but instead reminds us of the broader aspect of structure: the *interrelationship* of a number of components to achieve an entity.

This opens the door to a refreshing new dimension of looking at packaging *structure* as interrelating with other packaging components, more specifically, the packaging graphics. While it is true that packaging graphics and structure have their own individually defined objectives, Webster's definition suggests that the *combination* of the graphics *and* structure achieves a more effective *whole,* the total package, helping the marketer to build equity that will grow and expand the brand.

Everything about the package plays a role in communicating product imagery to the consumer. The package *form* can communicate images that influence consumer perception, appeal to the consumer's emotions, and motivate desire for the product before the consumer ever reads the label or sees the actual product. In that way, *the package is the product.*

Packaging can create perceptions in many ways:

- Ice cream in a paperboard or plastic tub creates a different quality perception than ice cream in a rectangular folding carton.

- Rice in a carton creates an image that differs from that of rice in a film pouch.

- A short, squat beer bottle communicates a different type of beer than a tall, long-necked bottle or a proprietary bottle design.

- A lipstick displayed on a blister card creates a different imagery than a lipstick in a foil carton.

- A watch placed in a velvet lined box creates a different price/value perception than the same watch placed in a plastic container.

- A leather belt in a canvas pouch achieves a different mindset than a belt hanging from a display rack.

- Straight-sided wine bottles identify wines from France or Italy; squat, bell-shaped bottles represent some Portuguese wines; tall, tapered bottles suggest wines that come from German regions.

In this way, many packaging structures determine at a glance what the consumer expects its contents to be.

Many opportunities to create strong brand recognition through uniqueness in package shape and handling features are missed. Too often, the extra effort required to achieve this is dismissed as unnecessary and too expensive. This is perhaps understandable if viewed only in the light of the costs required for design, research, and equipment changes and of the risk connected with the introduction of a new packaging structure.

However, there are enough brands marketed in unique packaging structures that *have* achieved enviable market leadership to give marketers encouragement to weigh the potential assets versus the burden of expenditures. Just think of the unique package shapes of Toilet Duck toilet bowl cleaner, Odol mouthwash, Tanqueray gin, Listerine antiseptic, Benedictine liqueur, Perrier sparkling water— all package *structures* that are easily recognized even if their labels were removed from the packages and all leading brands in their categories (see Exhibit 1.1).

Courtesy of The Perrier Group of America, Inc., Schieffelin
& Somerset Co., and Düring A.G., CH-8108 Dallikon, Switzerland

Exhibit 1.1

Short of having a unique shape for the *entire* package, there are many opportunities for innovative structural *features* that boost the perceived value of the product. Examples include the Tylenol "Fast Cap," the pour spout on the Morton salt container, the screw cap on Tropicana juice containers, the zipper-type opening/closing devices on Sun-Sweet prunes and Sargento shredded cheeses, the Tang and Kool-Aid measuring canisters, to name just a few.

Thus, it is possible to manipulate the imagery and positioning of a *product* by selecting a packaging form, material, and features that will influence the perception of the product and thus directly influence the consumer's buying decision.

Such opportunities are not necessarily a blessing. It is very easy to succumb to the temptation of wanting to be different by selecting product forms and materials that, while unique or novel, will *contradict* your marketing strategy. This is particularly true with regard to price/value perception. If package form or material is selected that does not match the consumer's preconceived perception of your product, it could have the reverse effect to that intended by you.

For instance, packaging common household nails in foil cartons would very likely communicate the wrong price/value perception for such common products. Conversely, shrink-wrapping an expensive tool to a plain corrugated cardboard panel would negatively affect the perception of that product. The physical appearance of the package plays a major role in guiding the consumer's perception and product approval.

Unfortunately, preoccupation with the economics of packaging can lead to packaging structures that contribute little or nothing to communicating anything positive about the product. Gable-top milk cartons, for example, are all structurally alike, and while they quickly identify the product category in the dairy cabinet, they communicate nothing unique about any particular product. Following the success, several years ago, of the unique Wishbone bottle that mirrored the *contours* of a wishbone, virtually all salad dressing brands, including private labels, followed its example with wishbonelike bottle configurations. Thus wishbone-shaped bottles became a *commodity* for every brand of salad dressing, neutralizing their ability to differentiate themselves except through their label graphics.

The same goes for most canned products, which utilize standard shapes and sizes, relying entirely on the label graphics to communicate something about the product. Round can configurations are the norm in the United States. Experimentation with can shapes has been virtually abandoned here, while in other countries, such as Japan and Germany, unique can shapes continue to be explored.

Even the automotive industry, which continuously produces distinctive new forms for their cars, does not follow the same convictions when it comes to packaging their aftermarket products, such as motor oil and antifreeze. Most of these products are offered in the same or similar-looking containers by virtually every marketer of these products, sometimes with minor design variations that parade as proprietary innovations. Yet, this is not necessarily the sole doing of the marketer. A few years ago, BASF developed a unique new "no glug" bottle for its antifreeze line. After a short-lived introduction of the new bottle, BASF capitulated to the inflexibility of container manufacturers who resisted the inconvenience and cost connected with production-line modifications at their plastic-bottle manufacturing plants when most other marketers stuck with the standard bottle shapes.

Fruit juices and drinks have long been marketed in many different physical configurations in Europe, ranging from unique glass bottle shapes to square paper containers with metal ends and pour spouts. Only recently have U.S. beverage marketers started to appreciate that novel bottle configurations can distinguish them from competitive brands and give their products a unique position in their industry.

The leading proponent of this is, of course, Coca-Cola. After having been available for many years in the same generic plastic bottles and cans used by the entire U.S. soft drink industry, Coca-Cola's introduction of plastic bottles and "contoured" cans that mimic its world-renowned glass bottle catapulted Coca-Cola from a nose-to-nose race with Pepsi-Cola to regaining the unquestionable leadership in the beverage market (see Exhibit 1.2).

To escape from the trap of standardization or near-standardization of packaging for all but a few brands, such as Coca-Cola, the development of an appropriately designed packaging structure can sustain a positive brand image and in that way enhance the perception of your product. Reporting on the new

Courtesy of The Coca-Cola Company

Exhibit 1.2

Coca-Cola bottle, *I.D.* magazine cited consumer comments such as "beautiful form," a "sensual look and feel," and "the ultimate enjoyment." No question about it—*this package is the product.*

What Graphics Tell the Consumer About Your Product

If structural design has the ability of creating images that appeal to the consumer's emotions, *graphic design*—the visuals that decorate the surface of the package—has an even greater opportunity to encourage the purchase of your product. To do so, the packaging graphics must be based on a distinct positioning strategy for the product and project this strategy in the most forceful and comprehensible manner.

This is not an easy task for brand management and the design consultant who seek to appeal to the consumer in ways that satisfy both themselves and the sought-after purchaser of the product. If you would ask the average consumer what attracts him or her to a packaged product, the answer would probably lead to the assumption that an attractive picture of the product is all that is necessary. While this may be sufficient to sell some products, the task of attracting the consumer is often much more complicated. You must ask yourself the following questions:

- How can our graphics attract the consumer's eye at the point of purchase better than our competitors'?

- How can we make our package look uniquely different from our competitors'?

- How can graphics communicate our product benefits better?

- What can our package visually communicate about our product that our competitors haven't communicated on theirs already?

"Every drugstore and supermarket is filled with shelf after shelf of half-successful brands," say Al Ries and Jack Trout in their book *Marketing Warfare,* and they explain their impression that the positioning of many brands is not clearly focused. While this may or may not be true, the fact is that packaging graphics are often expected to rescue a weak brand or product position and to transform it into a product with a positive image. This is difficult, if not unrealistic. Even the most cleverly conceived packaging graphics cannot rescue a bad product from eventual failure. Packaging graphics may succeed in cajoling a consumer into buying a product the first time, but if the product does not live up to the promise communicated by the packaging graphics, the consumer will not buy that product again.

The opportunities to communicate product attributes through packaging graphics are almost limitless. Graphics are capable of communicating *informative* and *emotional* messages. Informative messages include the following (see Exhibit 1.3):

- brand identity

- product name

- product description

- flavor or variety identification

- attribute description

- benefit statements

- sell copy

- promotional messages

- usage directions

- cross-references to other products and product variations

- nutritional information (for food)

- warning or caution statements (for drugs and chemicals)

- size and contents

Beyond providing pure information, the *emotional* aspects of packaging graphics are more subliminal. They evolve from the styling of various graphic elements, including logo styling, copy styling, symbols, icons, colors, textures, photography, and illustrations.

Says Thomas Hine, author of *The Total Package,* "People are affected by packaging in a way they do not consciously understand." Packages contain information that "consists of words and numbers, directed to the rational mind, while other facets, consisting of shapes, colors, and graphic expressions, bypass the rational and appeal directly to the consumers' emotions."

Brand identity, product identity, and copy elements in packaging

Front panel schematic   **Back panel schematic**

Exhibit 1.3

What then are the key components of packaging graphics? What are the elements that the designer can utilize to communicate a positive image about the product?

What Brand Identity Tells the Consumer About Your Product

The brand name that identifies your brand and product on your package is responsible for creating memorability, building brand recognition and loyalty, and providing product information. Styling of the brand name in a unique manner is, therefore, of primary importance to the current and future well-being of your product. A uniquely styled brand identity creates a recognizable "signature" that, just like your own signature does among friends, creates recognition among consumers and enhances their familiarity with your products. A uniquely shaped signature is also referred to as the brand's *logo*.

The logo can take many forms. It can be based on the brand name in some sort of unique typographic format or a uniquely styled configuration of the corporate initials. The logo can also take the form of a symbol that has an association with the product or can simply be an abstract shape designed to achieve brand recall. A bold logo will communicate strength, masculinity, effectiveness. A cursive logo usually communicates elegance, lightness, femininity, fashion. An angled or script logo provides an image of casualness, fun, movement, entertainment.

Regardless of its format, it is important that the identity is visually unique and that it is able to communicate a positive image and help to build consumer confidence in your product. There are many well-known logos that serve as examples of logo alternatives, consisting of styles ranging from simple letter forms to elaborately styled icons and signatures (see Exhibit 1.4). For example:

- simple lettering—Tylenol
- modified lettering—Exxon
- stylized lettering—Heinz
- script lettering—Kellogg's

- company initials—GE

- representative symbols—Quaker

- personality symbols—Sun-Maid Growers

- abstract symbols—Mercedes-Benz

- symbol and signature combinations—CBS

Because the brand identity on your packages is so critical in communicating a positive image to the consumer, it is important to keep it as constant as possible. Whether the logo is a stylized name or a symbol, whether it identifies a single product or an extensive line of products, whether it is used as a master brand or a subbrand, it is advisable that the logo style and proportions are maintained on all packages, regardless of packaging form, shape, and size. Unfortunately, many companies do not follow this rule. They allow the proportions of the logo to be altered depending on the package proportions, on the vertical or horizontal orientation of the packages, or to accommodate variations in packaging graphics. These marketers do not realize that every time their identity

Exhibit 1.4

changes, even if only slightly, it dissipates brand recognition and little by little weakens the franchise of the product.

These are serious mistakes that should be avoided under all circumstances. Your brand identity should rank as the most important link in the chain of product communications. Whether used in connection with packaging, print ads, TV commercials, sales promotion, signage, or letterheads, modifying your logo opens up a Pandora's box of continuous distortions that will ultimately lead to confusion among consumers. For this reason, the importance of never compromising the identity of your brand, except under the most extenuating circumstances, should be self-evident.

Even the color of the brand identity is important and works best if kept constant. While exceptions may be unavoidable, such as when product varieties dictate the need for color differentiation, it is essential to maintain the consistency of your brand identity as much as possible and thus assure a positive association with your products.

What Copy Tells the Consumer About Your Product

Next in importance to brand identity on packaging are the verbal communication elements, *i.e.*, the words that appear on the packages identifying the product and various information about the product. The verbal communication elements are of critical importance because they are responsible for communicating specific information about the product and its attributes. Depending on package size, this must often be accomplished within extremely limited confines of the label or package proportions.

The package has only a few seconds in which to identify its contents. In those few seconds, the key messages on the package must be communicated to the consumer clearly and effectively: What kind of product does the package contain? What are the product's attributes? What should the consumer learn from the package to make the product desirable? If the package does not accomplish catching the consumer's attention and its copy does not clearly convey your message, the consumer may pass it by and select a competitive brand instead.

The styling of the words by the graphic designer can tell the consumer much about the product. Bold sans-serif typefaces can communicate strength of product performance. Serif lettering can convey high quality, while delicate script styles can suggest softness, femininity, discretion, elegance.

Every information element on the package has to be precisely targeted and presented in an easy-to-read manner to communicate the intended brand and product information. As an example, the Totes package in Exhibit 1.5 clearly identifies corporate

Courtesy of Totes, Inc.

Exhibit 1.5

endorsement, brand name, slogan, product description, and product benefits statement:

Corporate endorsement: Totes

Brand name: Big Top

Product description: Oversized umbrella

Benefits statement: Opens "Golf Umbrella Size" to cover two or more adults. Folds small for easy carrying and storage

Company slogan: Rain rolls right off (Scotchgard logo)

Even numbers can be important. The identification of weight, fluid contents, and product counts are all important to the consumer. These may or may not give your product an advantage over your competitor. If they are to your advantage, don't be shy; play them up. If they are not, there is no sense in your trying to hide them. Consumers are not easily deceived.

In addition to such package design elements as brand identity, product identification, and attribute statements, many copy segments on packages are regulated by federal laws and sometimes by industry controls, such as

- usage copy to instruct the buyer how to hold, open, dispense, assemble, or store the package or the product within

- nutritional copy for food and beverages to guide the consumer regarding dietary concerns

- directions, indications, warnings, and dosage instructions for pharmaceutical products to ensure that they are used properly

- storage instructions, transport instructions, and various warnings for chemical products to avoid accidents

- contents statements, such as net weight, fluid ounces, and piece count for products that are bought by weight, size, or numerical preference

These copy elements are often difficult to accommodate in the small space available on the package or label, but they need to be strictly adhered to in terms of text, size of typography, and placement. Design consultants will be familiar with the most common of these, but mandatory regulations are so diverse and so category- and product-specific that a detailed review of these with the design consultant is always advisable.

Whatever your needs and requirements for packaging copy may be, it is prudent to be guided by the following considerations:

Keep it simple. Determine your priorities and resist the tempta- tion to tell the consumer *everything* about your product. A package is not an ad. Remember that the purpose of the package is to achieve instantaneous eye contact with a hurried consumer and doing so more quickly and effectively than competitive neighbors on the shelf.

Evaluate what is really important and what isn't. Identify the *real* benefits of your product and be sure that your package com- municates these quickly and clearly.

Prioritize copy elements on your package carefully. Determine precisely what is most important to communicate, what is sec- ondary, and what is less important to the consumer.

Keep it short. The fewer words, the better—as long as the words are carefully and precisely targeted to communicate infor- mation that will interpret your brand's position and convey its product's attributes.

What Color Tells the Consumer About Your Product

There is no question that color is the most emotional and subjec- tive issue in package design. We are passionate about color. All of us make color decisions every day of our lives. Whether it is the color of our house, a garment, a tie, a purse, a car, a fabric, or our stationery, we make conscious color choices that, in most instances, do not require approval or acquiescence from anyone (except per- haps your spouse) nor are subjected to (banish the thought) *research*. So why should our decision process be different regard- ing packaging colors?

Unfortunately, colors on packages that need to relate to a wide variety of consumers cannot be so easily dealt with. Color on packaging, whether a background color covering part or all of a package, color that identifies a flavor or draws attention to a product feature, or the colors of photographs and illustrations, plays a part in the image communication of your product. For that reason, color cannot be based simply on personal preferences, because the color perceptions of the consumers are too diverse.

It is difficult to hypothesize on the influence of any particular colors on product perception, although a few color pundits have tried to do just that. Nevertheless, some generalizations may help to understand the importance of color on packages, both in intellectual and emotional context.

- Color can identify a brand.

- Color can set a mood, such as fun, elegance, playfulness, or warmth.

- Bright colors tend to communicate lightness, festivity, relaxation, and joy, while darker, richer colors suggest a more serious frame of mind.

- Color can identify the color of the product inside the package.

- Color can assist in differentiating products, product varieties, and flavors.

- Green, a color unacceptable for food packaging a few years ago, is now a standard color for health-oriented products of *any* brand.

- Bright, lively colors are often used on cereal packages, because cereals are usually consumed in the morning, a time of day associated with brightness.

- White or light-colored packages communicate product attributes such as diet, light, salt-free, and low-calorie.

- Deep, rich colors, on the other hand, are often used for gourmet food and confections to communicate good taste, warmth, and appetite appeal.

- Gray and black on packages containing products such as cameras or electronic products reflect the "high-tech" colors of those products.

- White backgrounds on pharmaceutical packages suggest the efficacy of prescription drugs.

- Pastel shades, as well as black and gold, are often utilized on packages associated with fashion and elegance.

- Metallic foil, when used for visual effects rather than product protection, is almost entirely reserved for products that wish to communicate upscale and high-quality imagery, especially in packaging of cosmetics, gourmet food, and luxury products.

A more utilitarian use of color applies to product categorization and differentiation. Consumers have become so accustomed to color cues that identify certain product categories, flavors, sexual orientation, quality associations, and so on, that they respond to these cues almost automatically. An example of this is the colors used by soft drinks, where red cans and labels signify most colas, green stands for ginger ale, yellow for tonic water, and blue for seltzer. While it is *not impossible* to break out of this color syndrome, you have to be prepared for a rough ride. If you don't believe this to be true, just try to market ginger ale in a red can.

The most direct application of color is, of course, its use in representing the colors of the products inside the packages. For example, colored pencils, pigments, cosmetics, printing inks, and paints are usually identified through packages or labels on which at least a portion of the graphics relate to the colors of the products themselves.

Nevertheless, color is probably the one package design component that defies generalization more than any other. This is because the proliferation of possible colors, color combinations, and color shades provide a virtually unlimited color palette. The skillful use of these are, therefore, one of the most opportunistic tools for communicating product imagery.

What Pictures Tell the Consumer About Your Product

One of the *most* effective means of communicating product information and imagery is, of course, the use of pictures on the package. Photographs and illustrations on packaging identify products, describe their use, make them desirable, or create an emotional response by the consumer to the product inside. Photographs and illustrations on packaging are powerful design tools for

- *identifying product differences* or suggesting the end usage
- *communicating product functions*, such as describing step-by-step assembly of a modular product or procedures for applying a fixing compound or preparing a meal
- *adding emotional appeal* to a gift item, such as showing beautiful flowers to enhance the imagery of a gift item
- *showing the end result of using the product in the package,* such as a beautiful cake made from the flour contained in the package, a toy assembled from a construction kit, or the appearance of a room after the product in the package has been placed or installed there
- *imparting emotional imagery* by creating, for example, a feeling of speed (a runner) or relaxation (a sunrise), even though the product in the package has no direct relationship to such visual portrayal

Photographs or illustrations require particular and specialized skills, involving a number of exacting disciplines. They can be used in many different ways to appeal to reason or emotions in us. Photographs and illustrations are usually not interchangeable. What can be best achieved through photographs may not be obtainable through hand-rendered illustrations—and vice versa.

If photography is the medium of choice, the talent and experience of the photographer, the selection of models, the skill of food and fashion stylists, the arrangement of the subjects, and the lighting of them all can create substantially different visual and emotional reactions. All have a direct influence on how the product is perceived by the consumer.

Illustrators have similar opportunities of creating different images and emotional responses, but their work differs technically from that of photographers and is used for different reasons. While the camera records only what it sees through its photographic lens, the illustrator is able to manipulate visual images and interpret these in a variety of ways and by means of different rendering techniques.

The choice whether to use one or the other usually depends on the best judgment of the designer and on a number of circumstances, such as the subject matter, the product appearance, the product size, the intended message, the cost, the printing method, and the production timing.

Examples, to mention just a couple, include the following:

- For food, photography is unquestionably the favorable medium. An experienced food photographer with the help of a good food stylist can make the food on the package look mouthwateringly tempting. But avoid overpromising. Creating photographs that exaggerate the quality of the food too much or show it in an overly fancy environment may result in a disappointed customer who may not purchase the product again.

- There are instances when the nature of a product can be better and more effectively interpreted through a skillfully rendered illustration or a decorative interpretation of the product than through photography. For example, small objects, such as tablets or capsules on pharmaceutical packages, are usually more effectively handled through illustrations. This is because the tiny tablets often have surface imperfections that are magnified by the uncompromising camera lens and might require extensive and expensive retouching. A skillful illustrator can present the tiny tablet just as realistically as photographs but minus the blemishes.

When all is said and done, the package design elements—brand identity, product identification, copy, color, and pictures—linked to work in behalf of communicating product information and attributes to the consumer can be a powerful component of your marketing strategy. It is in your and the designer's power to use these elements to maximize a positive response from the consumer and thus affirm that *your package is your product.*

2 Package Design: More than a Pretty Picture

Package design is not like painting a landscape, a portrait, or an abstract expression of the artist's imagination. With the exception of a very small number of packages that have been selected by certain museums entirely on the basis of esthetic judgments, packages are not museum material. Their environment is not the quiet, leisurely pace of the museum, where exhibits appear several feet apart from each other and where strolling visitors study, admire, or criticize them according to their personal views and tastes.

In real life, packages are viewed in the midst of thousands of others competing for attention, shoulder to shoulder in supermarkets, mass-merchandise outlets, or department stores, where hurried shoppers glance at them for seconds if at all. Packaging, though charged with creating emotional appeal, has only one purpose: *to sell the product inside the package.*

To accomplish this, it is necessary for the designer to approach package design in the broadest possible terms. While some package design assignments may involve a single unit on which marketer and designer can lavish time and love, most packaging programs are more extensive and must deal with a whole host of strategic issues that go far beyond esthetic consideration. This does not mean that packaging cannot be esthetically appealing. On the contrary, the purpose of acquiring the services of a design consultant is, as the name implies, to make the packages as visually appealing as possible.

How can this be achieved? How can the package designer bridge the gap between stringent business requirements and the need for emotional appeal to attract consumers? Fortunately, this *can* be accomplished, as many packages that are both attractive *and* successful affirm.

But don't expect package design to materialize in a moment of creative genius. If only we could follow in the footsteps of Gary Oldman as Ludwig van Beethoven in the film *Immortal Beloved*, who appeared to create the "Ode to Joy," a Beethoven *icon*, in a flash of artistic brilliance. It just doesn't happen that easily—neither in musical composition nor in package design. As long as it is understood that the primary purpose of package design is not to create *a pretty picture* but to transpose strategic objectives into salable entities, the ability and experience that the design consultant brings to the table will result in packages that are *both* powerful *and* pleasing to the eye. Let's examine some of the key *nonesthetic* issues that enter into the package design development process.

Creating Brand Personality

One of the most important objectives in designing packages for new products or revitalizing existing ones is the creation of a *brand personality*. What is a *brand personality* and why is it so important to the marketing of products that are available under the brand's umbrella? Why is it a key issue in package design?

Walter Landor, the late pioneer of package design as a marketing tool, once explained it this way:

> Products are created in a factory, but brands are created in the mind.

If we agree with Walter Landor's theory, it follows that the *brand personality* is what creates an image of the brand and the product in the consumer's mind. Like a friend whom you love, a neighbor whom you dislike, or someone about whom you feel indifferent, packaging establishes images by which the products inside the packages are judged.

Think of personalities such as Churchill, Rembrandt, Reagan, DiMaggio, Hitler, or Marilyn Monroe. Whatever your opinion of these individuals, their names immediately conjure up pictures of them in your mind. You can apply the same to *brand personalities*. Some brands have such unique personalities that the mere mention of their names will immediately bring their *products* into sharp focus.

Who can ignore the powerful, sustained *personalities* of such brands as McDonalds, Oreos, IBM, Ivory, or Polaroid. Born as long as fifty to one hundred years ago, these brands have succeeded in establishing indelible images in the minds of consumers. And think of Levi's, Nike, Mercedes-Benz, Macintosh, Sony, and Häagen-Dazs. Each owns a memorable brand identity, a *brand personality,* that is instantly recognizable. You may like some of these brands (and use them) or you may not (and not use them), but you recognize them no matter how you feel about them. Without any doubt, these brands are top of the mind!

This enviable asset is not something that can be achieved in a sudden flash of brilliant thinking. These companies have carefully developed and nurtured their *personalities* by leveraging and updating their *brand identities* year after year. Indeed, your brand, and everything that symbolizes it, is your single most valuable asset. The image of your brand, the *personality* of your brand, can make an indelible mark on the consumer's mind. Without a unique and memorable brand personality your product may be worthless. Creating and maintaining a strong and lasting brand personality is at the core of package design and thus is one of the keys to marketing your products.

Brand personality results from the consumer's perception of a *combination* of elements that are linked and *gradually* mature into an effective and memorable brand image. These elements include the product itself, advertising, sales promotions, packaging, and, most important, the brand name and its visual presentation. This *gradual evolvement,* starting with conception of a clearly defined marketing strategy and evolving into a memorable and sustained brand personality, will eventually lead to instant brand recognition and, in turn, to brand leadership.

Brand identity is not necessarily just a name. There are *numerous* other means of creating brand identity and brand memorability. In addition to achieving product imagery, as discussed in the previous chapter, a memorable color, a package shape or feature, a symbol or logo, or a combination of any of these can create a recognizable brand personality. Many brands and their products can be identified instantly by their visual cues without the consumer's even reading their brand name:

- the yellow color of Kodak film packages

- the copper tops on Duracell batteries

- the pyramid-shaped graphic on the Marlboro cigarettes package

- the silhouettes of Perrier, Chanel, and Coca-Cola bottles

- the swath on Nike shoes

- the "snow caps" on the tips of Montblanc pens

These and many other brands have successfully developed unique visual identities. Whether through colors, icons, or packaging structures, they represent their *brand personalities.* Because they are maintained and promoted consistently year after year, it is impossible to gauge the value of these unique identities in terms of dollars and of consumer good will for their products.

The Importance of Brand Identity and Brand Equity in Packaging

If we consider Walter Landor's assertion, "Brands are created in the mind," the role of brand identity applied to consumer goods and their packages becomes increasingly comprehensible. Brand identity on packaging can play a decisive role in achieving a positive frame of mind regarding your products (see Exhibit 2.1).

Recent marketing conditions have accelerated the need for memorable brand identity. While in the 1980s an ever growing number of new brands appeared on the shelves, the economic slowdown of later years reduced this trend substantially. Advertising and promotional expenditures that were needed to launch new brands and relaunch existing ones came under increasing scrutiny by marketers who discovered that building upon the equity of an existing core brand through packaging was more effective, less risky, and less costly.

As packaging took on an ever greater responsibility in the distribution of products, marketers learned to understand more and more the critical importance of sustaining their brand identity. The brand identity is not only an important component that appears on the package, but its consistent use promotes recognition, suggests consistency and reliability, and invokes a feeling of confidence among consumers. This feeling of confidence in the brand's products is particularly crucial when planning brand extensions or the introduction of a new brand or product under an existing brand name.

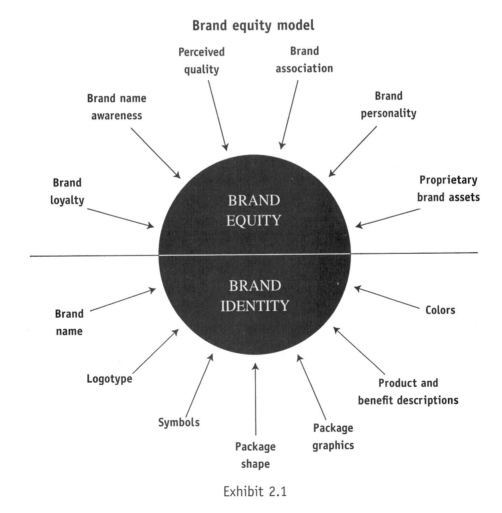

Exhibit 2.1

Let's look at an example. When Nabisco introduced Teddy Grahams, a new line of cookies, based on their considerable experience in graham cracker snacks, the products were developed to merge two specific branding strategies:

1. leveraging the health image of Nabisco's graham crackers, which have less sugar than cookies and thus appeal to mothers, who *purchase* the products

2. relaunching the core graham cracker product slightly sweetened and in bite-sized shapes to appeal to children who *eat* them

By combining the equity of Nabisco brands, well known and highly regarded by everyone (including children) with the goodness image of graham crackers among older consumers (mothers), this dual strategy paid off handsomely. More than $150 million worth of Teddy Grahams were sold during the first year after their launch—and they are still going strong today (see Exhibit 2.2).

None of this was easy. Extensive exploration with numerous images for the shapes of the cookies, such as letters, numbers, stars,

Courtesy of Nabisco, Inc.

Exhibit 2.2

sports, toys, and various animals, all researched in focus groups with children, pinpointed a clear winner. But without the powerful equity of the Nabisco brand identity, carefully nurtured over many years, the success of Teddy Grahams, as well as other well-known Nabisco products, would not be possible, regardless of how much creative effort was expended on new product and packaging concepts.

As the Nabisco example demonstrates, consistently used brand identification, applied to packaging as well as other media, is of *key significance* for marketing the brand's products and sustaining the brand's personality and recognition by the consumer. Brand identification on packaging is based on three key components:

1. brand frame

2. category dynamics

3. brand equity

Brand frame refers to the thorough, in-depth understanding of the brand's shopping environment and how your product functions in it. Now, you may think, this is so obvious; why even mention it? But the fact is that unless brand managers frequently visit stores to survey the state of their product category (and we are not sure that most brand managers do this regularly), they tend to have only a narrow picture of how their products look and how effectively they function in the real shopping environment.

As we discussed before, to know your retail environment is more important today than ever before. The dynamics of the retail trade are changing constantly and rapidly. Supermarkets are fighting to sustain their business against the onslaught of discount stores and many types of mega-outlets. There are substantial differences in how a product line functions in each of these environments, and the importance of staying current cannot be overestimated.

It is not only important to understand consumer behavior but especially helpful to be familiar with retailer attitudes:

- How do consumers shop the category?

- How does the retailer want the consumer to shop the category?

- What does the retailer think about your packaging?

- What about your packages is most meaningful to the retailer?

- Does the retailer have suggestions that could make your packages more effective in the retail environment?

Take a page from the revitalization of a line of business papers by Georgia-Pacific. The package redesign program for the product line responded to retailer and consumer needs in the office superstore environment, where numerous products are piled from floor to ceiling, labeling is confusing, differences between the technology of various papers are unclear, and there is little help from store personnel.

During pre-design focus group interviews with consumers and retailers, their many questions emphasized their confusion:

"Which paper will work best on ink-jet printers? Laser printers?"

"What does special ink-jet or laser paper do that regular copy paper doesn't? In what way do they differ?"

"How about colored papers—will they work on all types of printers?"

These and many other comments and an analysis of the display conditions at office superstores provided Georgia-Pacific with the opportunity of responding with a comprehensive label redesign program that accomplished two major objectives (see Exhibit 2.3):

- identifying Georgia-Pacific as an established and reliable brand

- clearly explaining the attributes of each type of paper

This example demonstrates how a thorough understanding of the *brand frame* can be vital in establishing the foundation for effective brand identity and package design if the brand hopes to gain consumer appreciation and category leadership.

Category dynamics is another important issue that affects brand identification and package design development. *Category dynamics* refers to activities related to the competitive array in a given product category as well as to current and anticipated category trends. Your product array could be affected not only by

Courtesy of Georgia-Pacific Corporation

Exhibit 2.3

directly competitive brands and products but by brands and products in unrelated categories. Fully understanding the category dynamics, many of which relate to changes in consumer lifestyles, is often overlooked and oversimplified by brand management.

Take, for example, the low-calorie food entree category. It was once totally dominated by Weight Watchers, but now the brand is fighting for its very life. Based on promoting low calories and less sodium, the products had a consumer perception of being bland and lacking variety—and their packages amplified this image by looking equally bland. As category dynamics shifted from "thin is in" to "staying healthy," consumers were seeking products that were *tasty* as well as beneficial to health. They turned to newcomers in the category, such as Healthy Choice and Lean Cuisine, while Weight Watchers, slow to react to the shift in consumer lifestyles, was belatedly forced to rethink its packaging to better communicate appetite appeal.

Similarly affected by lifestyle trends is the coffee category. The growing diversity of packaged coffee brands and flavors is born of the need to compete with not only other coffee brands in the supermarket but with the popularity of coffee bars, such as Starbucks.

These not only serve a wide variety of coffees at their stores but package and market them as private brands. Similarly, the dynamics of the beverage category have changed dramatically. Soft drinks that once dominated the category are now competing fiercely with a plethora of fruit juices, "New Age" beverages, flavored teas, and a growing horde of natural spring and mineral waters.

As consumers' tastes are constantly changing, brand managers need to be aware of shifts in *category dynamics* and be willing to modify their packages appropriately if they don't want to find themselves suddenly encircled by competitive and near-competitive brands and products.

Brand equity, the third important component in brand identity development and usage, is the most complicated. Brand management must have a thorough grasp of what the equities of a brand *really* are—not what you and your associates *assume* they are—and whether or not these equities contribute to a sufficiently positive image in the consumer's mind. If you do not have information that is current, say a year old or less, it is important to undertake consumer research to review the effectiveness of your brand identification.

Research can help in determining consumers' attitudes toward the category in general and your brand in particular. It can peel away the surface of your brand's personality as perceived by the consumer and probe which features of your brand, your product, and your packaging contribute to or detract from a positive brand image.

Brand equity is not necessarily the result of any particular feature of your brand or product. Brand equity is often born of the *accumulation* of brand imagery, brand recognition, brand personality, product usage, consumer consensus, and a variety of cues that trigger recognition by the consumer. Cues can consist of

- a brand name—Advil

- a trademark—Shell oil

- a product feature—Lipton's Flo-Thru tea bag

- a package shape—Michelob's bottle

- a package color—Campbell's soups' red-and-white label

- a symbol—Apple computers

- an icon—Pillsbury doughboy
- a slogan—Nike's "Just do it"

Once brand equity has been achieved, the vital importance of carefully nurturing this hard-won franchise cannot be emphasized enough. Brand managers who understand this will let no opportunity pass without emphasizing and reemphasizing their identity on packaging and all other media. Apple computers doesn't miss a beat in applying the apple symbol over and over again, from equipment to brochures and packaging. If the symbol cannot be reproduced in the well-known rainbow colors, such as on corrugated cartons, it will appear as a black silhouette, always reminding consumers of the fine reputation and engineering skills of that manufacturer.

The Lipton logo, recently redesigned for worldwide implementation, appears not only on all Lipton packages and in advertising but on the tags of each tea bag. Thus, hanging over the side of the tea cup, Lipton's brand identity is seen and recognized every day by millions of users in hotels, in restaurants, and at home (see Exhibit 2.4).

Courtesy of Lipton

Exhibit 2.4

This kind of marketing single-mindedness is key to achieving and maintaining brand recognition and thus safeguarding the brand's precious equities.

Constructing a Brand Architecture

It is not unusual in today's dynamic and ever changing market climate that a brand becomes so successful that it grows far beyond its original conception. There are many examples of this, such as Healthy Choice, Rubbermaid, Nike, Compac, Ralph Lauren, and Wal-Mart to name a few. When this occurs, it can be either a blessing or a curse: a blessing if management has the foresight to approach the growth pattern in a gradual and well-organized manner; a curse if the brand grows so rapidly that it progressively deviates from its original strategic course to a point where it becomes unmanageable and, as a result, unprofitable.

Unfortunately, the latter experience is the one that is more often encountered by marketers whose original elation about their success is mitigated by the realization that their pussycat has grown into a tiger. The number of SKUs spirals and line extensions are hurriedly and haphazardly launched, often overextending the brand. Intentionally or not, the once successful and well-managed brand of related products progressively becomes an unwieldy *megabrand* of numerous fractionated product categories.

Packaging, following the same track, magnifies the dilemma through a mishmash of brand and product identities that reflect the views of various brand managers instead of creating a cohesive brand program. The brand is in disarray, consumers are confused, and retailers are complaining. What to do? How to regain order without sacrificing the dynamics of the brand?

British philosopher Alfred North Whitehead put it this way:

> The art of progress is to preserve order amid change
> and to preserve change amid order.

The first step to preserve order is not to start redesigning packages but, as in any marketing endeavor, to develop a strategic plan. If the brand has outgrown its current brand strategy, a strategy that takes all the new factors into consideration must be conceived. When the objective is to reorganize an existing brand (as opposed to relaunching it to improve its consumer perception), the initial decision must deal with what type of overall organizational arrangement will lead to the most advantageous *brand architecture.*

Brand architecture establishes the strategic relationship between all elements of a brand line. This relationship will drive the design of the brand identity as well as packaging graphics for all products. Brand architecture could be one of several strategic alternatives. Among the most frequently used types of brand architecture are the following:

brand architecture	description	example
superbrand	a brand using the *same* brand identification and visual look for *all* products and packages	Duracell
leveraged superbrand	a brand consisting of several product lines that share a common brand identifier	Macintosh computers
brand endorsement	a brand with products that are marketed under separate sub-brand names endorsed by a superbrand identification	Nabisco
hybrid brand identity	a brand that mixes brand identification strategies, such as a brand that uses a superbrand identity for some products while subbranding or endorsing others	Ralph Lauren Polo by Ralph Lauren

The best way to visualize the brand architecture is to create an organizational chart that divides the brand into designated product segments and assign each product to one of the segments. Such a chart is very similar to the organizational chart of a large company that identifies various departments and assigns responsibilities to individuals in each department. Once established and agreed upon, the brand architecture is then able to function as the foundation that supports multiple levels of products in an organized and structured manner (see Exhibit 2.5).

Brand identification architecture model

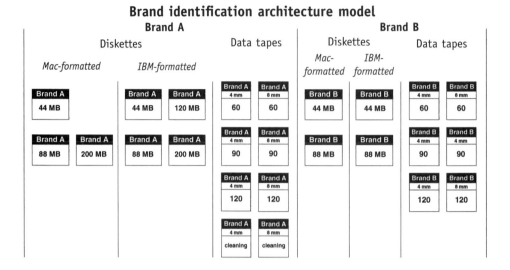

Exhibit 2.5

Brand Extensions and Package Design

In the current economic climate throughout the world, extending major brand names is a hot topic. When a successful brand has reached maturity, the desire of brand management is, of course, to seek new avenues for broadening the consumer base, leveraging the core brand, and achieving higher profits.

There are many good reasons for brand extensions. Today's consumers are so saturated with brands and products that many of them appreciate the simplicity of buying one brand to cover one or more product sectors. Even if each of those products is not the very best in its category, the consumer tends to feel that the whole is greater than the sum of parts. This applies to products marketed by companies in specific product categories such as the products of Kellogg's, Nike, IBM, and Rubbermaid, as well as products marketed by a company in a variety of categories, such as Pepsico (Pepsi-Cola and Frito-Lay).

Successful long-range management of core brands often hinges on a logical brand extension strategy. In order to make intelligent decisions, you must weigh the relative importance of several brand and marketing issues:

1. accurately identifying the category's image association
2. identifying products that could be linked to this association
3. selecting the best candidates from an available product array

But before marketing managers even consider extending a brand and modifying packaging accordingly, they need to analyze carefully the elasticity of the core brand and the benefits or possible damage that could result from packaging changes needed to accommodate the brand extensions. To achieve their goal without compromising the core brand, it is important, once again, for brand management to be accurate in the identification of consumer segments that comprise a relevant market and to determine any variables in advance of developing brand line extensions. Brand management needs to ask the following questions:

- How good is the fit?
- What is the brand's elasticity?
- What is the risk of diluting or changing the meaning of the core brand if packaging needs to be modified?
- Can packages of the core brand be adapted or modified to accommodate brand extensions without damaging the brand's equity?

Ideally, package design should be considered by brand management in the earliest development stages of the brand extension strategy. Marketers will usually benefit from such early input, enabling them to evaluate the potential effectiveness of packaging in propelling the brand extension into the market. When the package design strategy is developed in tandem with other strategic considerations, the launch of the brand extension has a better chance of success.

A technique that some companies use to determine the effectiveness of extending the core brand and how to best accomplish communicating the assets of the brand extension candidates is to brainstorm a wide spectrum of package design concepts in advance of making any final determinations. This technique is very similar to the storyboards and concept boards used by advertising agencies to communicate ad concepts to their clients. Like these, in order to cover the widest possible range in the shortest possible time, the package design concept explorations are most often just rough idea renderings. But what happens is that *visualizing* the brand extension alternatives, rather than just talking about them, results in a clearer understanding and often leads in unique new directions that were never anticipated when the brand extensions were first discussed.

After all is said and done about various marketing and package design issues essential to retail brands—brand personality, brand identity, brand equity, brand architecture, and brand line extensions—they all have one thing in common: while they have little, if any, dependence on package design *esthetics,* they play crucial roles in the marketing process.

Marketers who want to be winners in today's competitive world have come to realize that package design cannot be effectively consummated by their intuitive decisions alone, that in the package design process there are numerous strategic, intellectual, and emotional issues that play a vital role and that should not be minimized, because *package design is more than a pretty picture.*

3 The Strategic Role of Package Design

Packaging is a multibillion-dollar industry, one of the largest in the world. The typical supermarket displays twenty to thirty thousand products, and other outlets, such as department stores, mass merchandisers, and specialty stores, contribute to an ever growing mass of packaged consumer goods. A world *without packaging* is almost unthinkable. But this phenomenon does not come without serious concerns for the marketer.

For consumer products in particular, packaging can make the difference between success or failure in the marketplace and thus by inference the success or failure of the marketer responsible for the product inside the package. Marketers spend huge amounts of money to establish brand images through advertising, promotion, and merchandising. Then, at the critical point of purchase, when the consumer makes a decision in only a few seconds, the package must support that image and guide the buyer's heart, eyes, and hands toward the product on the shelf. The package operates as the final and crucial link in the communications chain from manufacturer to consumer.

The right package at the right time for the right product will boost sales and profits, while the wrong package can cause even a good product to fail.

The Decision to Design

Marketing strategy and market conditions are the fuel that drives the decisions regarding package design. The design process is far too complex, costly, and critical to the success of any brand or product—old or new—to be determined by a seat-of-the-pants attitude or to justify developing or changing a package merely for the sake of personal satisfaction.

For the consumer, *the package represents the product* at the point of purchase. It's a "friend" to whom the user has become emotionally attached. It's recognizable and comfortable, and it's a statement of brand values.

Changing the package can be risky because it can alter the consumer's perception of the product itself. Unless the marketing strategy is able to communicate that the product has been improved or has added benefits, redoing the package can cause the consumer to conclude that it doesn't taste the same or it doesn't work as well as before even when what's inside the package is identical to what was in it before it was altered. We know from consumer research and testing that changing the packaging graphics or altering the shape, size, or structure of the container can affect consumer perception of the product—positively or negatively.

How do you determine when the packages of a product or product line need redesign? That's one of the most difficult decisions for all brand and product managers. The risks involved in changing packages are substantial, and how to deal with them will be discussed in greater detail throughout this book. Most important, the decision for redesigning existing packaging must always be related to strategic objectives and should be based on a thorough comprehension of current marketing conditions in a given product category. Toward this end, research can often help in identifying consumer perceptions and attitudes towards the product, point out possible problems, and assist in making the best possible judgment as to whether or not your packages are in need of updating or other types of changes.

Whether and when to redesign is therefore one of the most sensitive issues facing the marketer. It is more likely today that a package will undergo some form of change than a decade ago. Swifter and more dramatic changes in the marketplace drive competitive conditions that must be countered. Because there are more and more products on the shelves fighting for a share of the market and the consumer's attention, changes in marketing strategy are inevitable. These often translate to changes in packaging. Among the most frequent reasons for considering redesign of an existing package or line of packages are

- product improvements
- competitive activities in the category
- falling share of market
- copycat competition from private labels
- changes in pricing
- changes in brand or corporate strategy
- additions to the product line
- new uses for the product
- changes in product weight or quantity
- availability of new packaging materials and containers

Each situation brings its own set of challenges, opportunities, and risks, and the marketing objectives for each situation can differ substantially. Let's examine some of these.

Package Redesign to Counter Competitive Pressure

Most packages undergo subtle changes over a period of time to maintain a contemporary look. These changes are most often *evolutionary,* designed to strengthen the brand's appeal without jeopardizing the consumer's favorable perception of the product or recognition of the package. There may only be a slight change in copy emphasis or a modification in information that may not even be noticed by the consumer.

However, the challenges from competitive marketers frequently leave marketing managers little choice but to consider changing their packaging. This becomes especially critical when competition comes from the very sources that distribute your products: the retailers. More and more supermarket and drugstore chains and mass-merchandising operators are developing their own proprietary labels that more often than not blatantly copy the leading brands. This is particularly evident in the over-the-counter (OTC) pharmaceutical and health care product categories. No sooner has a leading brand developed a new trade dress than it will find itself shoulder to shoulder with look-alike private labels on the store shelves.

This is a serious burden for national brands. The retailers thumb their noses at the brands' marketers, daring them to sue and thereby risk losing the retailers' goodwill. Only a few brand marketers with sufficient clout, such as Coca-Cola and Procter & Gamble, have actually sued *and won* legal battles for infringement of trade dress by some retailers. However, when massive legal funds for such lawsuits are not available to the marketer, the alternative choice is clear: instead of fighting the retailer in what could be an extended battle, redesigning the packages may be a more effective and expedient way of staying one step ahead of the copycat competition, at least for the initial period.

Whatever difficulties the pressure of competitive activities may pose, the decision to redesign should always be based on strategic initiatives and a thorough understanding of the strengths and weaknesses of your packages and those of your competition. This will often require consumer research to better understand the equities of the current packages. Which of these equities could or should be changed? Which should be retained? Does the consumer respond favorably to the brand logo or symbol? Or to the colors? Or to the product presentations? Or to the end-use benefits as described in the copy? How does the product or line of products measure up to the competition in terms of visibility in the retail environment? Are the colors and format recessive? Are they too much like those of the competition? These studies can be conducted in the stores or in controlled situations.

Marketers must understand that the responsibility for package design does not rest on the designer alone. Ultimately, the function of package design *is a function of marketing,* and therefore packaging must always be treated as a critical component *of marketing strategy.* For that reason, before any package design is undertaken, the right kind of information needs to be meticulously gathered and carefully analyzed in terms of competitive circumstances. Then the information must be clearly articulated in a set of criteria that will guide the design process.

Package Design to Update or Reposition an Existing Brand

The greatest challenge for marketer and designer is updating or repositioning a product, broadening its appeal to a wider audience while retaining loyal current users.

Brands are very much like people—they grow older, their surroundings change, the people with whom they interface change, their individual personalities change over time, and they sometimes develop problems that need correcting. Marketers must develop skills for recognizing the impact of each of these potential problems on their brand and make decisions on how to deal with them.

A brand or product line that has grown old can be either an asset or a liability for a marketer. Some packages seem to survive the changes of time and marketing techniques gracefully. Packages for such products as Cheerios cereal, Ritz crackers, Morton salt, Hershey's chocolate bars, Arm & Hammer baking soda, Budweiser beer, and Kodak film, among many others, have undergone such gradual changes over the years that few consumers, if any, have been aware of them. Notice how the "Morton Salt Girl" has been gradually modernized between 1911, when it first appeared on a Morton Salt package, and today, as shown in Exhibit 3.1.

But there are equally as many examples where major changes to long-existing packaging lines have been spectacularly successful. Among these are the dramatic switch of Breyers's ice cream packages from their traditional white to black backgrounds with mouth-watering ice cream photographs (see Exhibit 3.2) and Coca-Cola's launch of new plastic bottles and contoured cans that mimic the brand's world-renowned glass bottles. These are but a few examples of a company achieving indisputable leadership by making courageous package design decisions.

There are also horror stories of packaging changes that had disastrous results. The familiar attempt to update the packages of

Courtesy Morton International, Inc.

Exhibit 3.1

Camel cigarettes with more contemporary graphics several years ago met with strong resistance and nearly led to the brand's demise. When the company realized that they were off course, they did a complete turnaround, reintroducing the package with graphics that closely resembled the old, familiar Camel look.

Another contemporary impact on marketing retail products is the need to constantly *update packages in relation to new category situations*. Here again, strategy-driven objectives and an understanding of current market conditions are critical to initiating package design. There are substantial changes in the way products are marketed these days that effect package design in one way or another. Mass-merchandising outlets, warehouse stores, office superstores, drug superstores, and other sales giants are quickly changing the merchandising landscape. Supermarkets and department stores are constantly updating their marketing methods and shopping environments to compete with the mass merchandisers. Small specialty outlets and mom-and-pop stores are increasingly vanishing.

Responding to this merchandising revolution, packaging is trying to keep in step with the selling environments. Thus, Campbell's soup, long a holdout for its traditional stark red-and-white trade dress, has finally succumbed to the practice of virtually all food marketers

Courtesy of Good Humor-Breyers Ice Cream

Exhibit 3.2

for decades—that of featuring appetizing food illustrations on some of their labels. To retain the equity of its long-established brand imagery, Campbell's is careful in maintaining the familiar Campbell's script logo, the red-and-white label colors, and even the gold medallion as a link to its traditional imagery (see Exhibit 3.3).

As important as traditional values have been to some brands, marketers are aware that in today's fast moving economy, it is important for packaging to *stay in touch with changing lifestyles*. Today's young consumers, consisting more and more of working couples, are action oriented, far less conventional, and less product loyal than their parents were. Products are changing more frequently than ever before, and new products are being introduced that support the demands and lifestyles of the new generations.

Packages must be designed to fit these lifestyles. They must be easier to open and close, dispense their products more quickly, preserve them longer, and be more informative. Brand and product varieties are expanding at a dizzying pace. Brands that once were synonymous with specific categories suddenly overlap with brand categories with which they had no prior connection. Thus, Healthy Choice, once just a brand of frozen dinners, can now be found in almost every food product segment in the supermarket.

Courtesy of Campbell's Soup Company

Exhibit 3.3

Sony, once primarily known as a line of television products, now markets a wide range of electronics, from telephones, computers, and camcorders to music discs, all products that are part of contemporary lifestyles. Packages that contain and display these products at self-service outlets must be designed to appeal to the worldwide web generation and convey messages that communicate their end-use attributes quickly and effectively. More and more products make life easier, are more entertaining, and appeal to a more mobile and less patient consumer profile.

For some categories, *product personality* is particularly important. This applies especially to categories that appeal to the senses, such as cosmetics and perfumes. While the reputation and recognition of brands such as Chanel perfume, with its familiar square bottle and diamond-shaped stopper, is so solid that redesigning the famous package would most likely be detrimental to this brand, other cosmetic marketers, such as Revlon or Avon, have hectic programs of frequent packaging changes and new product introductions in packages meant to appeal to the emotions, some of them blatantly suggestive of sexual imagery (see Exhibit 3.4).

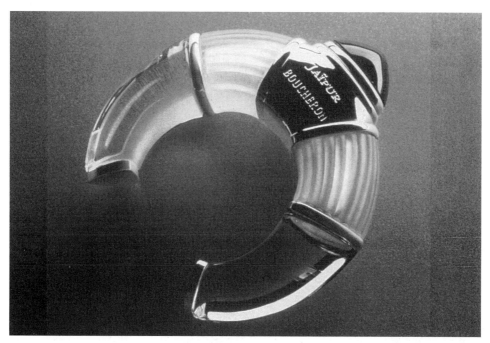

Courtesy Boucheron. Design Firm: Desgrippes Gobé & Associates. Creative Director: Joël Desgrippes/Sophie Fahri. Design Director: Corinne Restrepo.

Exhibit 3.4

Diametrically opposite to the beauty care category are brands and products with low levels of consumer interest, such as certain paper and household goods. These compete on a wholly different playing field and have great difficulty in achieving distinctly unique personalities. They are constantly challenged by competition, by private labels, and by the ever changing lifestyles of the consumers. Brands in these categories require sharply defined strategies if they want to explore new and unique ways of setting themselves apart from competition. Day-Timer is a good example of this. A leading marketer of personal organizer systems that were packaged in blank gray chipboard boxes and available only through mail order, Day-Timer's marketing team recognized the need for making their products available at office superstores. Toward this goal, Day-Timer created an entirely new line of packages and point-of-purchase displays. Instead of the plain chipboard cartons, Day-Timer packages now feature glossy photographs of the portfolios that glamorize their appearance and attributes and back panels that cross-reference to their large array of accessories (see Exhibit 3.5).

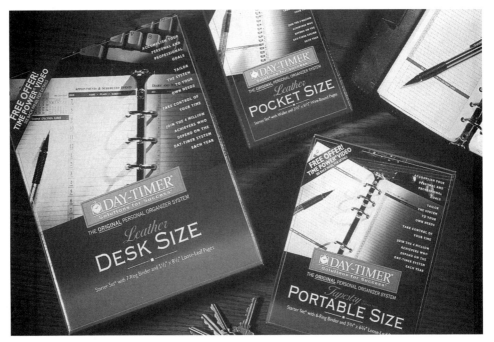

Courtesy of Day-Timer Concepts, Inc.

Exhibit 3.5

Package Design for New Products

The almost 90 percent failure rate of new product introductions has never deterred companies from developing and launching new products. New products continue to fuel excitement in the marketplace and often impact on a company's image beyond the new products themselves. Just as company technicians are charged with developing new and better products, marketing managers must develop exciting ways of marketing them. Design professionals must translate the benefits of these products into packages that will entice consumers to try the new products.

A "new" product can describe a variety of circumstances:

- a new flavor or product variety
- an improvement on an existing product
- a product not previously marketed by a company
- an entirely new product category

If a new product is simply *a new flavor or product variety*, packaging it is, in most instances, a relatively simple task for the marketer and the package designer. The main objective here is to make sure that the consumer is aware of the existence of the additional new flavor or product variety. In most cases, this requires visual and verbal communication elements that draw attention to the new item, such as a temporary "burst" describing what is new about the product. In cases where product varieties or flavors are identified through color coding, such as soft drinks, a new color may be the most logical method of identifying the new product.

When a new product is *an improvement on an existing product,* it is the responsibility of packaging to communicate the new product benefits quickly and clearly. This requires the right words as much as unique visuals. Such has been the case with the introduction of several new potent acid-fighting remedies formerly available only by doctor's prescription. These products—Tagamet HB, Pepsid AC, Axid AR, and Zantac—were introduced as over-the-counter remedies almost simultaneously during 1995 and 1996 and competed fiercely with each other. At the same time, they needed to develop new terms, such as *Acid Reducer* and *Acid Controller,* to identify their unique medical attributes and their claim of superiority over regular antacids.

When a company well-known for marketing products in a certain category introduces a product *not previously marketed by the company*, the most important issue will be how to relate—or *not* relate—the package for the new product to the company's existing product mix. Some brands have developed such a solid reputation that when introducing a new product of a different kind, marketing managers will lean heavily on the high esteem such brands have established among consumers. They believe that a consistent brand identity for various product lines marketed by the company will benefit the new product introduction, and they may have a point. Witness the proliferation of a variety of successful product lines under the same brand umbrella such as by Quaker Oats (cereals, rice cakes, frozen waffles, granola bars), Arm & Hammer (baking soda, detergents, toothpaste), or Fuji (photographic film, cameras, videotapes, floppy disks) to name just a few of the more conservative marketers.

In sharp contrast, when planning to add a new Southwestern-style flavor for their familiar line of Doritos a few years ago, Frito-Lay opted for introducing the new flavor under an entirely new brand umbrella. Thus the Tostitos brand was born and became phenomenally successful on its own.

Occasionally, a *new* product establishes *an entirely new product category*. Here the challenge to marketers and package designers is especially critical. Recent examples of such new product categories include the "New Age" beverages—Snapple, Arizona, Twister—as well as the new phenomenon of microbrewed beers that seem to spring up almost daily with new brand names and packaging. Other examples of *new* product categories are the popular "healthy" foods, represented most notably by the explosion of the Healthy Choice brand line, the detergent concentrates, electronic products, and many more. With all of these, consumers have to be convinced and packaging has to succeed in communicating that the consumers should take a chance by trying something *new* with which they were previously unfamiliar.

At times, the best intention to create unique packaging is a goal that is difficult to implement because, in certain product categories, packaging is expected to conform to standard packaging forms by which the consumers recognize the products, such as milk containers and snack bags. Thus when all liquid detergents are packaged in plastic bottles with easy-grip handles, it would seem foolhardy to introduce such a product in a gable-top paperboard beverage container. One detergent manufacturer who tried this several years ago had almost-fatal results when a child accidentally ingested some of the product, mistaking the gable-top carton for a beverage container.

The choice of how to position the package for a *new* product is always a difficult one. The rise or fall of the new product may very well depend on *the strategic role that packaging plays* in positioning the product correctly.

Most consumers are not particularly adventurous. They have grown accustomed to associating certain products with specific types of packaging, and they are not about to try their favorite product in an unfamiliar new package. Just think of Heinz ketchup, one of the most beloved product icons among young or old, despite its package being singularly impractical, either refusing to release its contents or letting it gush out like Niagara Falls. Yet consumers have steadfastly resisted all attempts for improvements to the Heinz ketchup bottle, relinquishing only some territory to its plastic companion package.

Consumers have also made it abundantly clear that they have no tolerance for new containers that offer no obvious advantages over the familiar ones. When developing a new package, consumer research should probe for consumer acceptance of the new unit in order to minimize the risk of launching a package that the consumer may judge as inappropriate and ill-conceived.

At the same time, the possibility of developing a truly unique container that provides a higher level of end-use benefits should not be discounted. The marketing team can explore with the designers and packaging manufacturers opportunities for an innovative approach that can complement the attributes of the new product and communicate its special assets. If you can combine a new product with appropriate breakthrough packaging, then you will have gained a significant head start on the competition.

Package Design for Brand Line Extensions

Compared to designing packaging for new products, designing packaging when introducing additional varieties to an existing brand should be, as we stated previously, a simpler procedure. The key is *advance planning* at the time the packaging strategy is *originally* conceived.

If packaging is *carefully planned at its inception*, adding a flavor to the line can be relatively straightforward. Adding a flavor to a line of soups or beverages, adding another product to a line of electric drills, or adding a stronger version of an OTC medicine to an already existing product array often requires only evolutionary changes in graphics, copy, or both. Occasionally changes in package *form* may be required, if only in connection with offering the product in a variety of package sizes.

- Campbell's has been able to add flavors to their soup product line with relative ease by maintaining their red-and-white trade dress even when subbranding a new product category such as Healthy Request.

- Black & Decker frequently adds new products to their line of electric and battery-driven tools using packaging with product photographs and copy to describe the product's features while maintaining the overall look of its product line.

- Tylenol differentiates its initial product, regular-strength Tylenol, from several product varieties (extra-strength, extended-relief, children's, and so forth) and a number of product forms (tablets, capsules, "geltabs") through package colors and product illustrations.

- Coca-Cola, originally a single product in glass bottles, is now available in cans and in several sizes of plastic bottles in addition to glass and in six varieties, Coca-Cola Classic, Cherry Coke, Diet Coke, Caffeine-Free Coke, Caffeine-Free Diet Coke, and Diet Cherry Coke, each identified through package design variations.

Occasionally, the introduction of a *new* product variety may require more than a simple modification of the existing line of packages. This can occur for various reasons:

- when an existing brand line becomes too extended and difficult to manage, requiring a new or modified strategy, such as restructuring the architecture of the line into several distinct product line segments

- when a new product or product group was not anticipated at the original product launch and is now added to a line

- when the *new* product variety does not neatly fit into the existing product array yet could benefit from association with the brand

- when the *new* product line extension is substantially different from other products in the line

When any of these situations occur, the entire existing line of packages may have to be reviewed and possibly undergo more extensive restructuring.

This occurred when Nabisco introduced SnackWell's, its line of fat-free and reduced-fat cookies and crackers. Nabisco's strategic objective was to create a brand line of "healthy" products and to display these *together* in the snack foods section of the supermarket. This strategy contrasted Nabisco's usual methodology of marketing individual brands each with its own distinctive trade dress and packaging graphics (for example, Oreo and Ritz). By opting for putting the entire line of *different* products under a *single visual brand umbrella*, Nabisco's SnackWell's packages achieve a powerful billboard at the point of sale (see Exhibit 3.6).

Line extensions also provide an opportunity to *rethink the line itself*. When introducing new extensions is an opportune time to create excitement that will mask the phasing out of slow-moving varieties without losing valuable shelf facings or antagonizing loyal consumers. Look upon line extensions as an *opportunity* to evaluate the effectiveness of your original package design format. Has it become dated-looking or overextended? How does your line compare to that of the competition? Could your line become stronger by weeding out one or two losers? Consider the addition of line extensions or flanker brands as golden opportunities to create excitement among consumers and strengthen your relationship with retailers.

Courtesy of Nabisco, Inc.

Exhibit 3.6

Packaging Relating to Corporate Identity Changes

A company's brand/product endorsement policy dictates the extent to which a corporate name or identity system impacts on package design. When a brand changes its identity, the effect on packaging may be minimal, requiring only modifications on portions of the packages.

But when a company undergoes a corporate identity change, management seeks to use the change as an opportunity to build awareness for the organization's products and capabilities. This may include a need for translating the corporate identity changes into new packaging. Typically, the designer who created the corporate identity system will also implement this on the company's packaging. This was the case with the development of the new corporate name, identification, and packaging system for Imation, a new company created when 3M spun off its four imaging and information divisions (see Exhibit 3.7).

But depending on how the client company is structured, other designers may be brought in at various operating levels to adapt the new identity system to product packaging. This happens especially with extensive identification projects, where corporate identification ranges from TV advertising and printed media to retail packaging and other media. In such cases, the designers must clearly understand the ground rules and the extent that deviations from the corporate identity system are allowed.

There are, of course, always special situations that require exceptions to a corporate identity system. The guidelines in the corporate identity manual are a helpful source for package applications, but it may be necessary to specifically define design parameters in advance of any design adaptations. If strict adherence to the corporate identity manual is mandatory, your design consultant should be made aware of this. If flexibility is permitted in expressing visually and verbally the relationship between the parent, its brand, its products, and its packaging graphics, then the designer may be able to explore design relationships that offer a wider choice of design

Courtesy of Imation Corp.

Exhibit 3.7

options, so long as these will not denigrate the overall corporate identity guidelines.

Managing the Package Design Process

With all these complex issues at work, what then guides design or redesign decisions regarding your packages? There is no single or simple answer to this. Remember, responsible package design is *integral to marketing*. Before any package design is undertaken, the right kind of information needs to be gathered and carefully analyzed, carefully organized, and then clearly articulated in a set of precise and strategic marketing and design criteria that will serve as a blueprint for managing the design process.

How to achieve all of this will be discussed in the following chapters.

4 Taking the Bumps Out of the Road

Now then. Where and how do we start a package design program? Developing a successful package design program is like building a house. You start with asking yourself questions. Where do I want to live? What town or city? What street? Who will be my neighbors? How big a house do I want? How many rooms? What style? What architect? How much will it cost?

It's the same with packaging. There are many important decisions that must be considered before the actual design process begins. There are many ways to reach these decisions depending on the nature of the business and the characteristics of the marketer.

Choosing Your Plan of Action

Some companies favor a highly entrepreneurial, fast-moving business style. These marketers are less concerned with procedures and careful documentation than with generating excitement and quick, dynamic action. These companies feel that in order to make an immediate impact in the market, they must be risk takers. They apply this philosophy to package design just as they do to other business decisions.

Other companies follow package development procedures that revolve around careful planning and attention to every detail, extensive documentation of every step in the package development program, and providing the time required for accomplishing their strategic goals.

The intention of this chapter is not to take a position for or against the fast-moving, entrepreneurial approach and the slower, more deliberate approach to package design. The entrepreneurial approach may be very appropriate for some companies and their products, but it must be understood that some of the steps discussed in this chapter will certainly be bypassed if an accelerated road is taken. The more careful and deliberate planning of package design is unquestionably less risky and usually more productive in the long run.

If the deliberate approach to package design is your choice, that will involve the following steps (see Exhibit 4.1):

1. conducting a thorough analysis of your product category

2. analyzing your and your competitors' packaging

3. establishing clear brand positioning objectives

4. identifying the attributes of your product

5. determining communication priorities

6. establishing precisely targeted package design criteria

Most important is to obtain top management's agreement to proceed with package design development and to organize the internal team that will work with you in moving the project forward expeditiously and efficiently.

Let's examine the development steps in greater detail.

Conducting a Thorough Analysis of Your Product Category

A package that successfully supports marketing strategy is one where the category conditions are thoroughly understood and addressed. To achieve this, you and your designers must carefully analyze your marketing strategy in relation to existing and anticipated marketing conditions in your product category.

Brand identity and package design development model

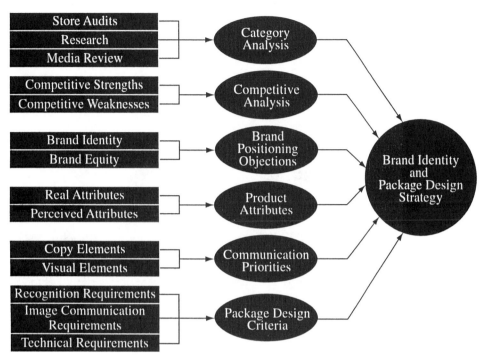

Exhibit 4.1

If research is available that is recent enough to be valid and that is targeted toward your product category, it is helpful to review it with the design consultant. Any other documents that shed light not only on your specific product category but also on related categories and competitive brands may contain valuable information in which the seeds for unique packaging concepts may be found. For example: if you plan to market a new soft drink targeted at teenagers, it will be useful not only to take a look at other soft drink packaging but also to examine packaging in other categories, such as snack chips and sporting goods, that target the teenager market.

While it is likely that the design consultants you will select will conduct their own research to uncover additional information about the product category in business publications, in libraries, or through professional information services, be sure to share with the design consultant any informational material already in your possession.

And don't forget advertising and promotional material. The consumer's image and perception of your brand and products is formed through the combination of all communication media. To be effective, packaging must operate as an integral link in the chain of communication elements so that the consumer will respond to your desired brand image, whether it is in TV commercials, print ads, point-of-purchase material—*or* packaging.

Analyzing Your and Your Competitors' Packaging

Knowing your opponent is always a key requirement in any adversarial encounter. Take your lesson from the supercompetitive world of sports. Baseball professionals carefully review taped replays to study the techniques of pitchers and others on the teams that they will face. The moves and strategic combinations of opposing football teams are given close scrutiny by their respective coaches before each game. Boxers use the first round to spar, trying to discover weaknesses in their rivals.

This need for knowing the strengths and weaknesses of your opponent applies to the development of package design if it is to function competitively at the point of sale. Nothing is as revealing as a visit to the stores in which your products are displayed next to those of your competitors. This is *where the rubber meets the road,* where your packages will either function effectively in your category in terms of findability and product communication or fail, giving your competitors an opportunity to beat you to the punch. So before you do anything else, conduct a category analysis by observing your packages as well as those of your competitors in the retail environment.

Just like in sports, observing your own packages and those of your competition on the real playing field—the stores where your products are sold—is the first important step in developing your package design strategy in relation to your marketing goals. Don't delegate this to others; go into the stores yourself. Even if you think that you know all that goes on in the category, we guarantee that you will find surprises. It will take the guesswork out of your package design planning.

It is also important that your category analysis is done with an open mind, without defensiveness regarding your current packages, even if you were responsible for developing them. Start analyzing the packages of your brand by asking the following critical questions:

- How easy is it to find your packages among competitive brands?

- Are your packages displayed the way you want them to be displayed?

- Do your packages function effectively in mass display?

- Are your packages hard to distinguish in the competitive array?

- Do your package dimensions represent your products most favorably?

- Do your packages open and close easily?

- Do your packages appear to be dated looking?

- Does your logo or signature clearly identify your brand?

- Is your brand clearly identified on all panels of your packages?

- Does your subbrand clarify its relationship to the core brand?

- Are the colors of your packages appropriate for the category?

- Do your colors clearly differentiate the products in your line?

- Does your package copy clearly communicate product attributes?

- Is the visual presentation of your products as effective as that of your competitors' packages?

Scrutinize your competitors' packages just as critically. Look for what you consider to be the strengths of their packages and what seem to be their weaknesses. In the event that you find your competitors' packages as effective as yours or even more effective, analyze the reason why. Does the *entire* package seem to be stronger than yours, or is there a specific part of the competitive package—logo, colors, text, dimensions, shape, or whatever—that achieves an advantage over yours?

Dissecting the competition will help you to better determine what to do about *your* packages. Should you leave well enough alone, or do your packages seem to need a change? If, based on your category analysis, improvements are indicated for your packages, the next question you have to ask yourself is how *much* should they change? Should the changes consist of an evolutionary update? Or is a stronger remedy required? Should you follow the precept "If it ain't broke, don't fix it," or should you risk shaking up the category with a dramatic package design change to ascertain your leadership position in the category?

In most instances, the answer to this rests somewhere between these two extremes. Understanding the range of options is critically important and will be discussed in detail later in this book.

Establishing Clear Brand Positioning Objectives

In his book *Building Strong Brands*, David Aaker, the well-known marketing strategist, defines brand positioning as "providing clear guidance to those implementing a communications program." *Position*, Aakers explains, is part of the identity that the brand must communicate. He describes *brand systems, brand leveraging,* and *consistency* as key elements for building strong brands. In all of this, the visual exposure at the point of sale clearly assigns to packaging a pivotal role for establishing and continuously communicating the strategy of your brand.

Before we discuss brand strategy communication on packaging, let's examine the significance of the brand itself. If we want to isolate the single most important element on a package, it would most likely be the brand *name*. The name of your brand is like the name of a person. The brand name identifies the brand, the product, or the line of products; creates memorability; and builds equity that will, in turn, build recognition and loyalty among consumers. The brand name thus becomes the cornerstone on which you can build and expand your strategic objectives.

But, warns Dr. Wolfgang Armbrecht, head of marketing communications of BMW, in a speech at the 1996 Corporate Image Workshop of the Conference Board, "An identity cannot be invented. . . . You cannot merely dream up your identity. It must come from the past and be inherent."

For this reason, the manner in which the brand identity is created and how it is presented and maintained over the years becomes a critical issue for package design. Your brand identity mirrors your brand's personality. It communicates reliability and value. It helps to remind the consumer to find and purchase familiar products, and, if used consistently and effectively, it helps to build confidence in your product, as well as products that may be added to the line at a later date under the same brand umbrella.

It is therefore critical for the package to be identified by a name, logo, or symbol that is unique, appropriate, and legally *ownable* (that is, not easily copied) and that communicates a positive and memorable image about your product.

Companies use many ways of identifying their brands on their packages. Electronic products marketers often identify their wares by using their corporate names, such as Sony and Xerox. Other manufacturers favor coined brand names for their products, such as Advil or Crest, with no reference to their manufacturers. Some favor more product-specific brand names, such as Healthy Choice or Sensodyne.

Conversely, the need for identifying a particular product quality or a distinct product feature frequently leads to a subbrand name that is used in combination with the master brand or primary name, often the name of the manufacturer. Subbrand names are intended to become memorable by combining a recognizable manufacturer's name that conveys confidence with a subbrand name that creates a product perception with which the consumer can identify, such as an emotion, an experience, a place, a description, or a perceived value.

There are many examples of subbrand names that have played significant roles in the marketing of certain products. Such well-known brands as Oldsmobile Cutlass, Johnson & Johnson Band-Aid, Nabisco Teddy Grahams, General Mills Cheerios, Exxon Superflo, and Gillette Sensor are but a few examples of successful and well-recognized subbranded products.

When subbrands are used on packaging, marketers must establish strict guidelines as to which brand element—master brand or subbrand—identifies brand strategy most clearly. Is the master brand name the most important element on the package, or is it the subbrand that the consumer recognizes? Should the master brand dominate on the package, such as with Hershey's milk chocolate, or should it endorse the better known subbrand, such as with Hershey's Kisses? If this important issue is not resolved for your package communications, the consumer will be confused and the equity of your master brand jeopardized.

Subbrand names can help to clarify important marketing issues. An example of this is when Coca-Cola made the decision to modify its popular flavor formula to better compete with the slightly sweeter Pepsi-Cola. When this resulted in loud outcries from loyal Coke users who wanted no part of the flavor change, Coca-Cola retreated to the original formula and made the ingenious decision of adding the subbrand name Classic to the traditional Coca-Cola signature. Thus, the initial blunder was turned into a brilliant marketing asset: Coke lovers immediately understood that Coca-Cola Classic meant the brand's return to the original flavor. So successfully has this maneuver been executed that the Classic designation survives today to differentiate Coca-Cola Classic, as the *regular* flavor, from the other Coca-Cola varieties.

Identifying the Attributes of Your Product

Who doesn't like to brag about his or her accomplishments? Products are no exception. All brand and product managers like to expound on the assets of their products—and why shouldn't they? They are expected to be proud of their products and tell the consumer about them.

But the relatively limited surfaces of most packages do not permit lengthy messages about product attributes. Packages are not ads. In packaging there is no opportunity, as there is in television commercials, for music and sound effects to set an emotional stage and thereby assist in communicating messages to the consumer. There are no special effects available for creating an environment or mood, no voice-overs to explain product features, no facial expressions made by talented actors to appeal to the consumer's emotions. Everything has to be condensed to fit the relatively small space on the package.

The consumer takes only a few seconds to identify the attributes of the package contents. If the package does not communicate the product benefits instantly, the sale may be lost. Thus, the general rule for identifying product attributes on packaging is the fewer elements, the better. *Keep it simple.* Clearly identify and prioritize copy and visual elements. In packaging, the slogan "Less is more" truly applies.

Determining Communication Priorities

One of the most perplexing and critical decisions regarding your brand strategy is how to establish verbal and visual priorities on your packages. What element on your package or label is the most important? Is it your brand name or your product descriptor? How about a unique attribute or a technical advantage? What is of secondary significance? And what is of tertiary importance? Or is the most consequential issue to present your product in a particularly attractive or appetizing manner through a photograph or an illustration?

The importance of being selective regarding copy or visual priorities cannot be overemphasized. Many times, multiple attributes all appear to be equally important. It's tough to make a decision about what to keep, what to put priority on, what to subdue, and what to discard. But those considerations force you to carefully sort out the order of priority of each element. Equal emphasis on too many packaging elements confuses the consumer and will not give the packages a chance to communicate effectively amid the cacophony of similar claims on competitive packages.

There is rarely space enough on packages to accommodate lengthy messages. In our fast-moving world of shopping, consumers do not appreciate or have the patience to read them all. A good package helps the consumer understand at a single glance what the *product* is all about. You must be a decision maker who has the courage to sort out what the most important messages are and to *delete* less important ones for the sake of maximizing communication clarity on your package.

Establishing Precisely Targeted Package Design Criteria

Based on your category analysis and retail audits, you should now have a clear picture of where your packages stand in relation to the category in general and your strategic objectives specifically. You can now develop criteria for your package design program, criteria that will guide the design process and help you evaluate design directions.

The development of package design criteria should be based on your marketing and package design brief, a detailed, *written* outline of everything that is relevant to the category and your strategic marketing plans. It is important to understand that developing a design brief is a way of thinking through, analyzing, and describing the entire process of marketing your product and outlining your objectives for designing or redesigning your packages. This will help in drawing attention to and clarifying any unresolved marketing issues before starting the package design process.

The design brief should be a detailed orientation explaining *all* the factors that feed into the marketing plans for your product and should address such subjects as

- history of the brand
- marketing background
- category conditions
- competitive products
- product varieties and the characteristics of each
- product attributes
- equity issues
- target audiences
- consumer age and income range
- current marketing trends in your category
- technical parameters
- cost and timing requirements

The information contained in the design brief will inform everyone involved in the package design process of your vision for marketing your product. Most important, the design brief will ascertain that everyone involved has the same information, avoiding potential misunderstandings and unexpected costs and delays.

In addition to providing a base of information, the design brief will lead to the formulation of the design *criteria*.

In a broad sense, the design *criteria* are to the design brief as the Ten Commandments are to the Bible. The Bible addresses a broad range of ethical issues by relating them to detailed descriptions of religious and historical events. The Ten Commandments *crystallize* these into a few brief, sharply focused ethical statements.

In the same sense, while marketing strategy establishes a broad spectrum of plans for your product, the design criteria should condense and *crystallize* these plans into a few sharply defined statements that will guide the development of package design directions.

Unlike the criteria for *marketing* strategy, which sets goals for the sale of your products, *package design* criteria outline objectives that specifically address the brand identity and package imagery issues.

Package design criteria are most often divided into three categories:

- recognition requirements
- image communication requirements
- technical requirements

Recognition requirements refer to those package design elements by which the consumer identifies the brand, product, or product manufacturer. These may be a corporate name, brand name or logo, product name, package color, package shape, or a variety of other elements that identify the brand and the product and that have become familiar to the consumer. For example:

- General Electric's widely recognized script logo has, through consistent use, gained such a solid equity position among consumers that it enables the company to apply it to products ranging from light switches to airplane jet engines.
- Johnson & Johnson's distinctive subbrand logo, Band-Aid, triggers instant recognition and product association.

- Colors, though less frequently utilized for brand identity, can spark instant brand recognition. The yellow Kodak film packages and the red-and-white Campbell's soup labels are examples of color fulfilling brand recognition requirements.

- Tanqueray gin is immediately recognizable by its distinctive bottle shape, as is the drop-shaped Perrier mineral water bottle and the legendary egg-shaped package for L'Eggs hosiery.

Image communication requirements are those package design elements that influence the consumer's image perception of the product. These may range from communicating the perception of the product's quality, its flavor, its efficacy, its efficiency, its gender relationship, or any number of characteristics about the product inside the package.

Pictures and copy can communicate product imagery; packing materials can communicate product quality, freshness, and product utilization; container shapes can communicate ease of handling, uniqueness, disposability, economy, and luxury. Every element and every component of the package participates in sending messages to the consumer's subconscious mind.

- Snack foods packaged in plastic bags communicate casual consumption, fun, entertainment.

- Products packaged in blister packs communicate ease of use, easy access.

- Cosmetics, by creating a myriad of unique bottle shapes, closures, and decorative visual elements, communicate imagery that appeals to the desire to be well groomed, attractive, sexy.

- Photographs and illustrations that communicate the appearance of the actual product in the package, sometimes glamorizing it, maximize appetite appeal or convey special features.

Technical requirements relate to package manufacturing. Technical information may include such data as

- information about plant equipment on which the packaging will be formed, filled, closed, and moved

- information regarding specific material requirements, especially for packaging of food, pharmaceuticals, and chemicals

- important details relating to product protection, such as product viscosity, odor prevention, product damagability, drop strength, shelf life, tamper resistance, and transport and storage requirements

- printing information such as substrates, number of available print stations, number of colors, color progression, types of coatings and finishes, and other production specifications

If a line of products uses a variety of packaging structures or if several suppliers are involved in the production of the packages, it is important that the technical parameters for each type of package and the production requirements of each supplier are identified.

Taking the Bumps Out of the Road

Package design criteria set the standard against which all package design development is evaluated throughout the design process. It is therefore critical for the success of a package design program that the design criteria for your packages address *specific* issues. It is not sufficient to rely on generalities. Objectives such as "We want to sell more products" or "The packages should appeal to men, women, and children of all ages" are meaningless and inadequate for guiding package design.

Developing precise and carefully weighted criteria, as described above—recognition, image communication, and technical requirements—is a key first step in creating the chain of communication objectives for package design. These criteria will serve as a constant pilot to keep the program on its course and save time and money. Neglecting or trivializing these important preparatory steps for package design will invariably result in lost time, unnecessary expenses, and frustration for all involved.

Let's look at a hypothetical example of design criteria for the package development for a line of "New Age" beverages.

Recognition requirements

- Identify all products as [brand name]
- Develop a bottle shape that is unique in the category and easily recognizable at retail

Image communication requirements

- Create a cohesive look for the entire line of products while leaving room for individual product personalization
- Clearly distinguish each product variety
- Maximize flavor communication
- Provide a color coding system for flavor differentiation
- Distinguish regular from diet beverages
- Appeal to young adults ages eighteen to twenty-four with medium income
- Achieve strong shelf impact

Technical requirements

- Wide-mouthed PET bottle, 36-millimeter finish, hot filled
- Plastic screw cap
- Maximum label area and a neck band
- Must run on existing filling equipment

If this were a real project, the design criteria above would become the focal point of all developmental steps that follow, as well as the yardstick against which all design concepts would be evaluated. Thus, it is paramount to understand that precisely targeted design criteria have two fundamental missions:

- to guide the design process
- to evaluate design directions that have been developed

It is also important that design criteria, together with all other information pertaining to the package design program, are confirmed *in writing* and agreed to by all involved in the package development. In the fast-moving world of marketing, personnel that initiate the package design program might be replaced at any time in the design development process. If communication objectives are put in *writing*, it will be easier for the new marketing team to become familiar with the strategic intent of their predecessors and to assess whether to continue, change, or modify the existing package design strategy or strike out in an entirely different direction.

With design criteria thus determined and agreed upon, the *bumps have been taken out of the road* and you are now ready for the next step: determining to *whom* you will entrust the creation of your package design.

5 Selecting the Designer

If you have ever been faced with selecting a professional consultant, be it a lawyer, financial advisor, architect, or computer programmer, you know that it can be a daunting but exciting experience. Selecting a designer or design consultant is no different. The selection process for finding the right consultant involves the same steps as any process that requires judgmental decisions. It involves a step-by-step procedure of comparing, sorting, and eliminating various available options and eventually making your selection.

What Is a Package Designer?

The package designer can bring to the table an unusual combination of skills and experiences. These include an in-depth understanding of a wide range of marketing- and packaging-related issues. Package design is based on a mixture of intellectual, creative, and technical components. These may include, in addition to three-dimensional design and graphic design capabilities, understanding strategic issues, product positioning, consumer research, sales, and merchandising, as well as technical aspects regarding prepress preparation, printing processes, packaging materials, packaging machinery . . . and the list goes on.

In addition to dealing with many of these issues on a daily basis, the package designer must stay current on a broad scope of information, from trends in consumer lifestyles and fashions to technical innovations, from mass distribution methods to the latest changes in government regulations pertaining to packaging.

The package designer may be called upon at any stage during the design development to apply this knowledge to solving a range of visual and technical problems unmatched in any other form of communication. This includes design, composition, logo development, color, lettering and typography, photographic techniques, and illustration styles, as well as a wide variety of printing- and production-related subjects. Dramatic advances have been made in computer technology; the designer must be fully conversant with electronic procedures for design and preproduction preparation.

Because the skill and experience of the designer must be so intellectually broad and technically proficient, the selection of the "right" designer for any given brand identity or package design assignment is critical for the project's success.

Searching for the "Right" Designer

The sources for brand identity and packaging design professionals are numerous, and wide differences in capability and experience among designers should be expected.

The selection process for a package designer or design consulting firm should be based on an evaluation of track record, reputation, breadth of service, business references, facilities, staff, and ability to service accounts.

The most critical criteria for selecting a designer or design consulting firm is to evaluate the design professional's ability to understand marketing and to develop unique brand identity and package design solutions that will support a given brand strategy. With these criteria in mind, the design consultant should be able to demonstrate his or her ability and experience in having solved strategy-related problems for other clients.

This experience may or may not include expertise in a specific product category. While category-specific experience can be a valuable asset in a design consultant, this need not necessarily be the *primary* selection criterion. In fact, it may be more advantageous for the marketer to select a consultant that can demonstrate strong capabilities in problem solving. By *not* having been deeply immersed in a particular product category, the designer, unhampered by judgments formed during previous exposure to the category, could act as a stimulus and bring a fresh viewpoint to the table.

When searching for a design consultant, ask yourself the following questions:

- Do the experience, capabilities, and services described by the design consultants' presentations and literature match their actual track record and professional reputation in the field?

- Are the design consultants you are interviewing truly experienced in solving complex strategic problems in a unique manner?

- Can they demonstrate past experience in skills and services particularly relevant to your needs?

- Do you prefer a relaxed, personal relationship with your design consultant, or do you value strictly professional and organizational assets? With this in mind, does the personality of one consultant fit your firm's culture and work methodologies better than another?

- Does your project require a consultancy that specializes in particular design or marketing skills, or are you looking for a firm that offers a range of services, such as consumer research or name development in addition to design?

- Are you looking for design consulting services for a specific assignment, or are you seeking a long-range relationship with the consultant?

- Does your strategy suggest a highly creative, cutting-edge design approach, or are you targeting a very conservative consumer?

- Does your budget allow you the freedom of looking for the best available design consulting source?

Sources for Brand Identity and Package Design

Design sources vary greatly, and all have advantages and disadvantages. Selecting the right design consultant is an exciting challenge that requires the marketer's best judgment as to the design skills and experience required, the complexity of the assignment, and the available budget.

Large design consulting offices usually offer a wider variety of experiences and a broader range of special skills, such as often required by large corporations. Fees charged by large brand identity and package design consultancies are sometimes but not always higher than those of smaller offices and may be based on their reputation and availability of personnel experienced in a wide range of disciplines. These may include services such as brand evaluation, consumer research, name development, and merchandising and technical expertise in addition to the expected design capabilities. The availability of these services may or may not provide added value depending on how critical such services are to the marketer's strategic objectives.

When selecting a major player in the consulting business, the marketer will usually interact with a number of people in that consulting office. The primary contact may be the firm's owner, its president, or a key executive and may periodically include individuals in creative or technical disciplines. Before making a final selection, the marketer should request meeting key members of the consultant's project team who will service the account.

Small consulting offices and individual design consultants exist in large numbers. Some of these have gained previous professional experience in the employ of larger design firms and have left these to open their own design consulting offices. They usually offer a moderate range of design services, often revolving around their specific design philosophies and styles. Some designers specialize in graphic design exclusively; others offer structural design and technical services exclusively. Few small design groups are proficient in both, though many claim they are. Small design con-

Thus, when seeking assistance in brand identity and package design development, the options boil down to the marketer's preference for professional and strategic experience, technical expertise, or personal service.

The Business of Design

It is important to realize that design consultancies are business enterprises just like any other. This means that professional designers are not just "creatives" who live in a world of their own. Quite the contrary, they understand financial goals and responsibilities and will work with you in a businesslike manner. The creative experience that design consultants contribute to the marketer's business is provided in exchange for specified fees, just as the marketer's products or services are offered in exchange for specified prices.

For that reason, the fees and related costs for a design assignment should be treated in much the same manner as setting the price structure for a line of products. To avoid the possibility of misunderstandings and to protect both the marketer (the buyer) and the designer (the seller) from later disagreements, it is important to both parties that a *written proposal* be submitted by the design consultant and approved by the marketer in advance of the start of any actual work. The proposal should clearly outline the procedure and methodology for the project. It should describe the design consultant's understanding of the assignment and delineate the work that the marketer will receive for an agreed-upon fee.

Most design consultants divide their proposals into several phases that will build on one another toward the final solution. The phase descriptions will identify and describe each step, such as marketing analysis, retail audits, consumer research, concept development, revisions, finalization, internal and client meetings, travel, supervision of and coordination with outside services, and such other activities as will be needed to accomplish the design program. Design consulting fees are usually based on the amount of staff time required to accomplish each step of the assignment and on the complexity of the assignment.

The proposal that a design consultant prepares should clearly and precisely identify all aspects of the design development program, including

- the procedure for handling the assignment
- time requirements for completing each phase
- fee requirements for each phase
- out-of-pocket expenses for each phase
- method of payments
- contractual obligations pertaining to the consulting services
- contractual obligations for subcontracts, if any
- design ownership provisions

In addition to the fees for the design consultant's services, the marketer will be responsible for out-of-pocket expenses for purchases made by the designer in connection with the assignment. Depending on the requirements of a given assignment, out-of-pocket expenses may include such items as travel, photography, models, typography, research facility fees, competitive products, shipping costs, and other purchases made in the marketer's behalf. Such out-of-pocket expenses will be invoiced in addition to service fees. Some design consultants bill out-of-pocket expenses at their actual cost; others add a small percentage as a service charge for handling and financing such purchases; others combine both methods.

In any event, it is always advisable for the marketer and the consultant to have a clear understanding as to the manner in which out-of-pocket expenses will be billed so that any misunderstandings at the time of invoicing can be avoided.

As in any business, the marketer has the option to negotiate the terms of the proposal with the design consultant (see Exhibit 5.1). Design consultants quote what they consider to be reasonable fees for the work outlined in their proposals. They rarely reduce fees on demand but are usually amenable to discuss fee modifications that are paralleled by procedural changes or an appropriate reduction of services. If budget restraints are a concern, the consultant may modify certain steps that are less crucial than originally anticipated; or it may be possible to reduce certain portions of the services or handle some of these through the company's internal resources. Conversely, if additional meetings or services are required for which the proposal did not provide, the proposal may need to be adjusted to fund these requests.

Four steps of selecting and negotiating with the design consultant

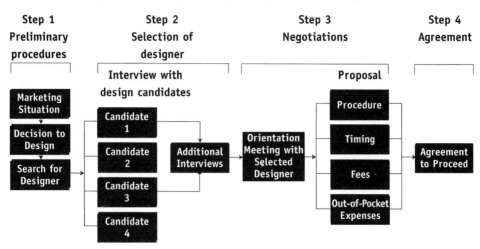

Exhibit 5.1

Design consultants understand that negotiations are part of their professional activities. There should not be any concern by the marketer regarding engaging the designer in budget discussions. If budgets are limited, designers usually encourage their clients to *negotiate* contractual modifications with them, rather than to solicit proposals from seemingly cheaper design sources.

Once the proposal is agreed upon and signed, it becomes a legal contract between your company and the consultant. The consultant as well as the company will be obligated to adhere to all the provisions outlined in the signed proposal. This means the following:

- The consultant is obligated to live by the terms of the proposal regarding services and fees described in the proposal. The consultant cannot and should not submit any additional fees for services that have not been previously approved by the marketer.

- On the other hand, any changes to the proposal that are initiated *by the marketer,* such as requests for additional services, changes in marketing objectives affecting the design direction, copy alterations, or accelerated artwork deliveries, may result in renegotiating one or several phases in the proposal.

- In the event that the design program is discontinued by the company, the company is obligated for payment to the consultant for work completed to date or as otherwise specified in the proposal. The terms of the proposal will outlive the departure, change of employment, change in responsibilities, or shift in the positions of any client personnel who initiated or approved the design assignment.

A word of advice: *any deviation* from the terms outlined in the proposal of which the marketer may not be aware—especially anything that may be interpreted by the consultant as *additional* services—should always be confirmed *in writing* to alert the marketer and avoid possible misunderstandings at the time of billing.

Some companies require that proposals by an outside consultant be submitted to their legal departments. In that event, negotiations may take place between the marketer's legal department and the consultant's attorney. While legal issues in design will be discussed in greater detail in a later chapter (Chapter 12, "Packaging and the Law"), it suffices to say at this point that *some* legal issues will always enter into negotiations regarding consulting services.

Working with the Design Consultant

Package design development programs are most often assigned on a project-by-project basis, although some companies prefer retainer arrangements with the consultant.

A retainer arrangement between a company and a consultant obligates the consultant to do a specified amount of work for a pre-arranged monthly or annual fee. It also obligates the consultant not to work for a company that markets products competitive to the marketer's products. In return, the marketer commits to giving preferential treatment to the consultant when assigning other design projects or even to using the retained consultant exclusively. Thus, the consultant enjoys a degree of security by being guaranteed a predetermined amount of income, and the marketer is assured of the consultant's undivided attention.

Some marketers prefer to shop around for a design consultant for every new assignment, presumably in search of design variety or lower costs, or both. While this may appear a logical business procedure, the on-again, off-again method of working with design consultants can impact negatively on the marketer's long-range strategic goals, since it precludes the benefits that accrue from a mutual understanding of the marketer's business and the consultant's methodology through a continuous working relationship.

Similarly, a methodology that can have a negative impact on the effectiveness of an assignment is when the employment of a design consultant is made dependent on the development of preliminary design concepts *without payment*. Such requests for *speculative design submissions* are sometimes offered to several design firms with the implied prospect of being awarded the full assignment if one of the free submissions is selected. While this may be a method acceptable to architects and ad agencies in competing for multimillion-dollar assignments, most brand identity and package design consulting offices consider *speculative design* contests as unrewarding, as well as not in the client's best interest, as free design precludes their investment in time for *in-depth investigation* of the category and a full understanding of the marketer's strategy.

Establishing Costs and Timetables

Once the marketer has made a decision as to the type of assistance needed for brand identity and package design development, regardless of whether using an internal company source or external consulting services, the next steps will be to establish specific procedures, allocate costs, and set up timetables for the project.

As previously discussed, just as the makeup and character of various design sources vary greatly, methodologies and costs for developing brand identity and package design programs also vary substantially.

If a company utilizes an *internal design department*, design costs incurred by the department are, most likely, accrued on an annual basis and amortized as part of the departmental or overall corporate budget. This may give the impression of the availability of a "free" service. However, more often than not, internal designers are required to estimate the cost for *each project* and charge these to the requesting department or to the brand for which design services are required. While this will not make the internal design department a free service, it can provide a less expensive design source, especially when the design department budget can be disbursed among several departments.

Advertising agencies, if directly involved in package design for a brand, will often allocate this cost as part of their annual service to their clients. However, in keeping with recent ad agency arrangements of working for their clients on a project-by-project basis, the agency may allocate package design development as a separate service. Some agencies will subcontract brand identity and package design to outside specialists or recommend their own affiliated design consulting services, if applicable. These services will be invoiced to the marketers either through the agency or by the design consulting office directly.

Design consultants will, of course, always invoice their fees and other expenses for design development to the marketer's company directly. What does this involve? How do design consultants arrive at their fees?

Where Do the Design Dollars Go?

The method and approach of estimating brand identity and package design fees vary substantially from one firm to another and appear to be a profound mystery to some marketers. Design consultants are often dismayed when asked to estimate the costs of a design program within the ubiquitous *ballpark* figure after being offered the scantiest of information about the product and the marketer's strategic objectives—even sometimes without any meaningful information at all.

The truth is that design budget estimates cannot be arrived at in a flash and in a vacuum of non-information any more than a marketing or advertising budget can miraculously arise from sudden inspiration. As with everything in the communications business, design costs vary with circumstances and are always based on elements of time, labor, experience, depth of service, and most important, the buyer's ability to purchase services the marketer can afford.

It is also important to understand that design costs can be affected by the *chemistry* between the marketer and the design consultant. Designers, no matter how capable, can only be as good as the way they interface with their client's staff. This is because more often than not the budget for brand identity and package design is not an isolated component, but almost always part of the marketer's *total marketing budget,* which may include costs for product development and production, sales, advertising, promotion, publicity, and test marketing. While brand identity and package design development services are proportionately minor cost elements, they are nevertheless part of the marketer's overall budget consideration.

Design fees are broken down into functional segments relating to the design requirements and the participation of various professional disciplines needed for a design project. Depending on whether the strategic objectives call for structural design, graphic design, or both; whether they include other requirements, such as category analysis or consumer research; how many SKUs are involved; and many other factors, design projects can range from a few hundred dollars for minor design modifications all the way to six-figure mega-budgets for extensive brand identity programs.

So, how can the marketer evaluate the costs for brand identity and package design development? It's not anything that one can touch or see or hear. What the design consultant contributes to the marketing communication process is a combination of ideas, thoughts, logic, and practical experience. But how much does one pay for thought and logic? How much is an idea worth? How many dollars cover practical experience?

With such ethereal matters to deal with and so many variations in the approach to design by various consultants, the best way to understand the cost elements of brand identity and package design is to hypothesize a "cost pie" divided into a number of "slices" in a range of sizes. Seeing each slice as representing a part of the whole process will be helpful in gaining a better understanding of *where the design dollars go* and thus will offer the marketer and the design consultant an opportunity to tailor the design program to the marketer's best advantage at an affordable price.

Assuming that a design program is divided into several basic consecutive steps, the *percentage* of the whole cost required to accomplish each step can guide the anticipated costs and can be used to tailor the design program by adding and/or curtailing cost elements to achieve the objectives without hurting the end results. The steps are as follows:

1. Orientation

2. Category analysis

3. Brand architecture development

4. Structural design

5. Graphic design

6. Consumer research

7. Structural and graphic refinements

8. Design finalization

Dividing the pie by percentages, the charts shown in Exhibit 5.2 will approximate the relationship of various cost elements. Keep in mind, however, that each marketing strategy has different requirements so that each will result in different slices and that each design consultant will very likely divide the "pie" differently.

One thing that the pie charts do not illustrate is the *hidden costs* of doing business. Because marketers usually meet only the design firm's principals or key executives, they tend to overlook that as in any business costs go beyond the actual design work. Design

Percentages of brand identity and package design development costs

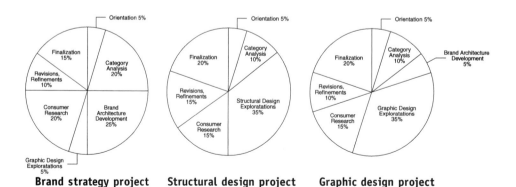

Brand strategy project   Structural design project   Graphic design project

Exhibit 5.2

offices, like any other business, must be properly equipped to function effectively. The efficiency of the office space and equipment—administrative and studio areas, conference facilities, computers and computer software, communication equipment, and other important tools of the trade—are expected by marketers to be state-of-the-art. Many of these, especially computer equipment and software, are expensive to maintain, and their costs increase every year. Thus, when reviewing a brand identity and package design proposal, all of these aspects must be kept in mind if the success of the project, the working relationship with the consultant, and the satisfaction of the marketer are to be realized.

Ready, Set, Go!

Once the fees and the methodology outlined in the proposal have been agreed upon and signed by the company, the consultant is ready to start the actual assignment. In most instances, a signed proposal automatically signals the beginning of the work outlined in the proposal.

Occasionally, however, because of pressure to meet a project deadline, such as the launch date of a product line, it may be difficult to delay the start of the project while waiting for the tedious approval process of some companies. If it is necessary for the consultant to start working on the project before the signed proposal is available, to circumvent the possibility of embarrassing misunderstandings, a *written letter of agreement* between the marketer and the design consultant should briefly outline their mutual obligations until the budget approval process has been accomplished.

When *selecting a designer*, the efforts expended during the negotiation process will go a long way to provide for a pleasurable and productive working relationship for both marketer and design consultant. The marketer's business will benefit from a lasting and trusting relationship with the consultant, and may be one of the smartest investments made toward achieving long-range strategic goals.

6 Preparing for the Race

The creative development of a brand identity and package design program is the most exciting, stimulating, and challenging part of the package design process. It is what many brand managers consider the fun part of their responsibilities. While it is true that design development should be fun, achieving the most effective and unique solution for implementing your brand strategy depends on a thorough understanding of the development process.

But where do we start? What is important to get the creative process going? What information does the design consultant need to proceed with the design project? What role does the ad agency play in the package development process? What is the best coordination procedure with the design consultant, the ad agency, and the marketer's production staff?

There are, of course, as many procedures as there are manufacturers, design consultants, and advertising agencies, and each will tout their own preferred methodology. Look at the creative development segment of a brand identity and package design program as if it were a relay race. You are at the starting line. You have selected a design firm; now you are ready to begin the first lap of the creative development process. Each lap will bring you closer to your goal of beating your competitors.

Steps in a Design Program

To prepare for the race and for the purpose of discussing how to manage a design program, let us assume our procedure would be based on the following step-by-step methodology (see Exhibit 6.1):

1. Marketing and category analysis

   - orientation meetings to review strategic issues and all factors pertinent to the brand identity and package design program

   - review of advertising and sales promotion plans

   - review of manufacturing and packaging facilities

   - pre-design research to understand consumer attitudes toward the brand and the category in general

   - review of package design criteria

   - category and marketing information

   - analysis of your packages and those of your competitors

2. Creative development

   - development of a broad range of brand identity and package design explorations

   - selection of concepts for further development and/or consumer feedback

3. Consumer feedback

   - focus group or one-on-one interviews with consumers

   - research analysis and selection of final concepts for further development

4. Modifications and refinements

   - design refinements and modifications of selected concepts

   - development of preliminary three-dimensional mock-ups

   - post-design consumer research to help select the final design

Step-by-step procedure of design development

Exhibit 6.1

5. Finalization and implementation

- final mock-ups or working models of new structures (if applicable)

- adaptation of the selected design to additional products, varieties, sizes, and packaging forms

- pre-production communication with package supplier and/or color separators and printers

- package design finalization (working drawings and/or mechanical art)

- production and/or printing follow-up

No matter what methodology is used, it should always begin with orientation meetings.

Orientation Meetings

Orientation meetings are the beginning of communications between the marketer's and the design consultant's project committees. The purpose of these meetings is to *orient* the design consultant's staff as to the objectives and parameters of the assignment and to familiarize the consultant's staff with strategic goals and other factors pertinent to the package development. This is when designers *listen* and have an opportunity to ask questions.

The more information given to the designers during the orientation meetings, the better. Not only do the designers benefit from this input, but the effectiveness of their solutions may depend on the quality of information that is given to them during the orientation meetings.

The information that the designers are looking for pertains to the history of your product, your marketing plans for the product, and advertising and sales promotion plans for your product.

History of the product that should be reviewed includes current or previous packages (or pictures of these), annual reports, and any other printed material containing historical data about the company and the company's brands and products.

Marketing plans, current and future, for the brand should be discussed freely and openly with the design consultant. It is important that the consultant be considered a trusted and intimate member of your marketing team, not a vendor. This is critical to the design consultant's ability to develop packaging that will not only satisfy current marketing needs but also provide for future marketing opportunities, such as brand line extensions.

Advertising and sales promotion such as television commercials, print ads, and promotional material relating to the brand should be discussed with the consultant. Since package design is a communication medium, TV storyboards, animatics, and/or preliminary ad layouts will help the consultant understand how advertising intends to communicate the brand position to the consumer.

The Marketer's Role as a Catalyst

Who should the marketer bring to the orientation meetings? As stated before, the information given to the design consultant (and staff) during the orientation period impacts on the project's objectives and the designers' ability to develop the most effective package design solutions. For this reason, all those responsible for package development and design approvals should be included in the orientation process. Especially important is that the orientation will include the key decision maker. Otherwise there is the danger that the key executive may not agree with the design recommendations, resulting in time-consuming and expensive revisions or even rejection of the concepts. Everyone will be frustrated and the project will suffer.

Making certain that everyone on the team is "on the same page" is the responsibility of the marketer who pays all of the bills and has the leverage. As marketing becomes more and more complex and package design projects involve vertically integrated layers, such as advertising, promotion, and merchandising, the package design consultant's solutions will be used by other consultants in their problem solving. Therefore, it is crucial that during the early stages of planning, the diverse talents that contribute to marketing the product are considered and that input is elicited from *all* of these sources. It is the marketer's responsibility to act as a catalyst to bring all of these players onto the team so as to create linkage that will maximize impact and sales power for the product in the retail environment.

The Linkage Between Advertising and Package Design

The ad agency is the most visible implementor of brand strategy. It is charged with the responsibility of creating brand imagery and communicating product attributes to the target audiences. When products succeed in the marketplace, the marketer is the hero. But when a product fails or loses share of market, the agency is often blamed. This puts the agency in a tenuous position. Agencies no longer can rely on their long-term associations with clients as they did in the years past. As advertising costs skyrocket, clients are switching agencies more frequently than ever before and seeking alternatives to strengthen brand and product communications by placing more reliance on influencing the consumer at the point of sale. Package designers are benefiting from these changes. Thus, it has become more critical to the marketing process than ever before that agency and marketer integrate the designer into the planning procedure as early as possible so that every member of the team starts from the same vantage point.

In the initial strategy meetings between marketer and agency, the design consultant should be included and thoroughly apprised regarding strategies and goals. This information will formulate

brand identity and package design criteria, provide design direction, and help shape design development of the packages. If the
challenge is to relaunch an established brand, the design consultant needs to know whether the advertising strategy will make new
claims for the product. If it concerns the launch of a new brand or
product line, it is even more important for the designer to understand how the ad agency intends to formulate the strategy and the
imagery for the new venture. Since packaging must support this
strategy over a relatively long period of time, the designer should
accommodate the needs of the advertising claims while developing
a package format that will operate successfully if and when the ad
strategy is modified or changed.

There will be times during the course of a project when the
marketer will work more closely with the ad agency and other times
when it will be the turn of the design firm or another outside consultant. However, the marketer should bring together *all* the parties whenever critical strategy or positioning issues are discussed
and decided (see Exhibit 6.2).

Levels of involvement

Exhibit 6.2

Coordinating the efforts by the advertising agency with those of the package designer, the merchandising and promotion specialists, and others involved in the image development process should not be viewed by any of the parties as a threat or dilution of status. The experienced package designer knows that an understanding of the marketer's brand positioning strategy and the agency's responsibility for transposing this strategy into exciting advertising is crucial to the formulation of design criteria and the productivity of the brand program.

Develop a Team Dialogue Early

Much of the misunderstanding between ad agency and designer can be traced to the differences in how each approaches design. The agency's creativity is based on a *campaign* mentality. Campaigns change from year to year, often from season to season, and, in some categories, even from month to month.

There is an important difference between advertising and package design. A package is more personal to the consumer than an ad. The package is held, touched, stored, used to dispense the product, looked at many times at home on a shelf or on a table. A package becomes part of the user's life, even a sort of friend, and may need to be recognized in this way for many years. In contrast, ad agency creatives, driven by immediacy, are constantly challenged to generate new ideas and to pack excitement and attention-getting visuals and copy within the time constraints of fifteen- or thirty-second TV commercials. The package designer, on the other hand, works with a medium that emphasizes long-range effectiveness that does not change with the same frequency as ads.

One of the more costly mistakes by marketers is bringing in the design firm late in the process. When a new brand is launched or an existing product repositioned, the marketer tends to turn first to the agency. The package designer rarely participates in the early stages. In fact, it is not unusual for the design consultant to be brought in *after* product strategy and positioning have been determined. Therefore, the designer plays catch-up when advertising and promotional plans are already well on the way to implementation. The design consultant is forced to digest important information in a relatively short time and late in the image development process for the brand.

If this happens, an opportunity may have been missed to gain the benefits of the designer's input and wealth of experience in brand image building. The most successful marketing programs are those in which the designer begins to function as a team member during the conceptualization phase. This is the period when strategies are explored and formulated. It's never too early to worry about package design. While it may be too soon to begin *designing* packages, it is not too soon to initiate discussions and planning for the *kinds* of containers that will give the product a real advantage in the marketplace and think about dispensing methods, closures, materials, shapes, and other features that will help the product stand out from the competition.

Without the marketer's encouragement, ad agencies tend to avoid involvement with the package design consultant, except for their concerns about how the package will look in a TV spot or in print. Conversely, there are times when the designer feels threatened by the agency's more influential relationship with the marketer. In some situations, the ad agency and designer may even find that they are competing in certain areas of design. Brand identity development, for instance, tends to reflect a short-term brand promotion mindset when agency designers take a stab at package design. Since the brand's visual identification is the most critical element in the marketing mix, brand identity specialists are more experienced in developing long-range imagery for packaging that withstands the fluctuations of advertising campaigns and that will maintain brand consistency for many years. Fortunately, this sort of conflict is diminishing as the stature of brand identity and package designers is rising among marketers and ad agencies.

Cooperation, like coordination, must be an ongoing mindset, not a sporadic happening. Open lines of communication among the team members will significantly enhance everyone's contribution. The agency may have a particular feel about the "attitude" of the marketer's product because it is devising the brand's imagery. Sharing this attitude with others involved in building the brand can prove invaluable not only to agency but to design consultant as package planning and design directions are explored.

Getting to Know Each Other

Team spirit among the ad agency, production personnel, researchers, suppliers, and the design specialist can build mutual rapport and respect through better understanding of each other's philosophy, experience, and practices. When the client assembles the project team, a number of steps should be taken to ensure smooth communication between *all* levels of operations, from the primary contact to the backup contact of the respective specialists.

Too often, our vision of others is colored or limited by our particular occupational skills. We may acquire a narrow point of view and miss out on opportunities to tap someone's know-how or experience. "Get to know you" visits should start early in the project. Each of the groups involved in the project—ad agency, design firm, researchers, display houses, and others—can benefit by making brief presentations of their capabilities to other members of the project teams. This will be time well spent, broadening understanding of the range of available resources and often sparking ideas that will be helpful to everyone on the team. There is no substitute for one-on-one contact. Different people's roles take on more importance as the design effort progresses. If any of the project participants need specific information, they should know whom to contact and feel comfortable about it. Learning about each other's professional capabilities will create mutual respect, make working together more pleasurable, and first and foremost, benefit their mutual client, the marketer.

Review of Package Design Criteria

Design criteria set the standards against which the package design concepts will be evaluated. It is, therefore, important that both marketer and designer agree on design criteria in advance of the design explorations. As we discussed in previous chapters, recognition requirements, image communication requirements, and technical requirements should be clearly outlined and discussed in great detail with the design consultant. In addition to reviewing brand positioning objectives, communication priorities, product attributes, and copy emphasis, conditions peculiar to the category that may impact on the design development need to be brought to the attention of the consultant and thoroughly discussed at an early stage of the design development program.

Pre-design Consumer Research

An in-depth understanding of consumer attitudes toward the category in general and the brand in particular is of special significance to the designer.

If category information is lacking and no recent research on packaging in the category exists, the design consultant may recommend pre-design consumer research to acquire such critical understanding. What interest the designers most are the following: what visual cues the consumer associates with the brand; what kind of product information is important to the consumer; what priority should be given to information on the package that may influence the purchase behavior of the consumers.

When redesigning the packages of an established brand, a loyal consumer base exists that must not be jeopardized. To product users, any change in *packaging* is often perceived as a *product* change. For this reason, it is critical that the designer distinguish between existing brand and packaging equities that should be retained and those that could or should be altered.

If attitudinal and behavioral information is not available, the designers will have to rely on past experience to make design decisions based on judgment alone. Past experience is certainly a valuable component in any development process. But because conditions in the category may have unexpectedly changed, the risk of making wrong decisions based on past experience is greater than when decisions are based on up-to-date consumer research.

Pre-design research, most often qualitative in nature, can probe crucial issues and produce directions that will guide the designers' creative development. While this type of research involves a limited number of interviews and, therefore, should not be considered as *definite*, valuable learning from the information obtained can help to broaden the search for design solutions.

Category and Marketing Information

In order for the design consultant to harmonize the marketing strategy with existing category conditions and achieve package designs that are uniquely different from the competition, the conditions within the product category must be fully understood.

Even if the consultant has had experience with the category through previous assignments, additional information, articles in the press that discuss the category, and any other sources of information can contribute to unique design solutions.

Consumer profile information such as age, sex, and income of the target consumer that will impact on purchasing patterns in the category is critically important. The design consultant should be apprised about the product's target audiences, demographics, and purchasing patterns—the kinds of people who are going to purchase the product and the kinds of people who are going to use the product. Is the product aimed at a particular market? Young or old? Male or female? Affluent or not? Is the target consumer both buyer and user? Many times, brands are not skewed to a single target audience but have a primary audience, such as mothers who purchase the product, and a secondary audience, their children, who are the ultimate consumers. Each of these must be addressed during the design development and therefore should be thoroughly reviewed during the orientation meetings.

Retail and lifestyle conditions often differ in various parts of the country and even more so in other countries where the brand may be marketed. Products that in Kansas City are neatly lined up on shelves may be found in dump displays in Los Angeles. Bostonians buy tea bags for hot tea, while in Atlanta iced tea predominates. Italian salad dressing, the favorite in New York, is less popular in Denver where ranch style dressing flies off the shelves. Analgesics, easily purchased off the shelves of any U.S. retail store, must be asked for in European pharmacies where they are kept behind the counter. All of these situations will have a substantial impact on the creative approach that your design consultant will take and, therefore, must be clearly understood.

A list of competitive products in their order of importance is crucial information for the design consultant. All major competitors should be discussed and reviewed in full detail. Competitive marketing activities and packaging should be rigorously evaluated. This is especially important if competitive packages are not easily available (for example, packages for medical products, prescription drugs, chemicals, high-tech electronic equipment, and so on) or if the products are available only in local or regional areas. The more you can communicate about competitive activities to the design consultant—especially the strengths and weaknesses of competitive products—the better informed the consultant will be.

Retail audits will take the design consultant's staff into the retail environment where current packages as well as those of competitors are displayed or are *expected* to be displayed. By going into the stores, the designers can become familiar with the retail conditions in the category, such as shelf heights, peg-boarding, dump displays, behind-the-counter storage, and a myriad of other conditions. Are products in the category displayed vertically, horizontally, on Peg-Boards, in dispensers, in corrugated shippers, at eye level, on top or bottom shelves, on special merchandising racks? Are they displayed together with other products marketed by the company or are they checkerboarded with competitive and private labels? Stocking and restocking practices should also be observed.

In the ever changing retail environment, a detailed discussion should include category situations, retailer attitudes toward the product, distribution problems, stocking and restocking conditions, price application methods, and so forth. While the consultant's staff will conduct their own retail audits at appropriate locations, the more the marketer can add to the consultant's understanding as to how the product is marketed and how it functions in the category, the better able the consultant will be to assist the marketer in developing meaningful design solutions. Remember, no matter how experienced a design consultant may be, the marketer is probably still better informed about his or her own brand and products and those of competitors than any consultant could ever be.

As part of the category analysis, the design consultant's staff photographs packages and competitive packages in the actual retail environment. These retail audits are reviewed at the design consultant's office to better understand the conditions under which the products are marketed and to evaluate the strengths and weaknesses of individual packages. Special attention is paid to shelf impact, brand identity, product presentation, copy, and overall design.

The photographs will be used as reference throughout the design development process and are among the most important and most informative parts of the design explorations. Retail audits should be conducted for *every* new design assignment. It is not sufficient to rely on retail audits that were done six months ago or even earlier, because the competitive situations in the category will most likely have changed during that period of time and could make a major difference in the approach to the design assignment.

Corporate or brand policies relating to your product or to all products marketed by the company should be identified to your design consultant. Otherwise the consultant may work in a vacuum that will, sooner or later, impede the package development process. Design development that will need to be altered in order to conform to the corporate and brand policies with which the consultant could not have been familiar will inevitably result in wasted time and money.

Review of Manufacturing and Packaging Facilities

Information regarding manufacturing and packaging lines at the company's production facilities is essential to the designer, regardless of whether this information involves graphic or structural packaging changes. If current packaging utilizes specific packaging machinery, assembly, and filling procedures, transportation and damage control methods that may affect the packages or the products, the designer should be advised about these in as much detail as possible.

A visit by the designers, the consultant's technicians, or both, to at least one manufacturing plant is an important prelude to a design program, particularly, of course, if the package design assignment entails structural package design development.

Your methods of manufacturing, filling, assembling, stacking, and transporting products may differ from the operational procedures of other manufacturers. Visiting a plant facility will familiarize the designer and the consultant's packaging technician with the limitations as well as opportunities for the package development. Even if the manufacturing and packaging facilities are not uniquely different from other similar facilities, a plant visit will assure the design team that no potential problems have been overlooked.

If a unique structural departure from your current packaging is planned, it is especially important for the designer and the consultant's technical staff to understand the packaging methods and equipment. This is particularly consequential when the objective is to launch the product in a unique, "out of the box" container that may require plant equipment modifications or even entirely new equipment. In that event, the associated costs and time requirements should be evaluated and weighed against anticipated benefits. It is also not unusual that the consultant's technical expertise may result in suggestions that will lead to cost *savings* in packaging structures and materials.

At the very minimum, a representative from plant production should attend the initial orientation meetings. This will familiarize plant production personnel with the *strategic* objectives for the package design program and encourage them to better understand package equipment modifications, should they be needed.

Scheduling Design Development

Last, but not least, is the importance to the consultant of clearly understanding the marketer's time requirements. These should be geared, first and foremost, to timing in connection with the product launch schedule, especially if such schedule is related to a specific season. Beyond that, however, schedules should also consider the design consultant's need for a reasonable period of time for creative development and finalization as well as time for consumer research, if needed, and for production requirements.

Finally, there is the need for meetings throughout the project. And who hasn't experienced frequent delays in getting all the appropriate individuals together for periodic review meetings?

Schedules should provide for realistic meeting dates, leaving some room for the inevitable postponements that are as frustrating as they are inescapable.

Now, all of this may seem a terribly lengthy and exhausting prelude to the actual design development process. Some may feel that the process discussed in this chapter overcomplicates what had been expected to be a simple procedure of assigning the design project to a designer who could whip up package designs in a couple of weeks.

But the importance of having a thorough and well-organized orientation program prior to proceeding with actual package design cannot be overemphasized. All professional consultants will agree that the more information they obtain at the beginning of the assignment, the better they are able to help marketers to achieve meaningful design solutions in the shortest period of time and in the most economical manner.

Thus, in *preparing for the race* the marketer is making an investment in the realization of the strategic objectives and the ultimate success of the brand identity and package design program.

7 Creative Development: Where the Rubber Meets the Road

We have now reached the last lap of our relay race. The baton has been passed to the design consultant for the final and decisive lap—the concept explorations.

The approach to concept explorations is likely to vary substantially from design firm to design firm. Some will approach the concept explorations with predetermined convictions and present a limited number of solutions or even a single design recommendation. Others will present a wide range of design concepts. While there is no right or wrong way of handling concept development, there are general guidelines to follow.

Depending on variables, such as the consumer's familiarity with the brand, the type of product, the number of products in the line, the length of time elapsed since the last design change, category characteristics, and marketing strategy, the explorations may range from design alternatives that are *close-in*—meaning that they are based on a close visual relationship to the existing packages—to more unique, "out of the box" concepts.

How the design assignment will be approached by the consultant will depend, to a great extent, on the magnitude of the project.

An extension to an existing line of products may require only a limited range of concept exploration since the current brand identity probably will be retained. There are numerous examples of line extensions where the overall look of the brand is affected minimally, if at all, such as the addition of a flavor to a soft drink line, a new fashion color for a series of lipsticks, a novel fragrance for a detergent, an improved version of a software system, or a new pattern for a line of bedsheets and pillowcases.

On the other hand, if the line extension to an existing brand will stretch the product lineup far beyond what was originally perceived, it may be necessary to rethink and possibly reconstruct the entire line, resulting in much more complicated design explorations.

This, for example, was the case when Frito-Lay planned to add a Mexican flavor to the existing line of Doritos brand corn chips. Believing that this popular megabrand had reached its saturation point with several well-established Doritos flavors, Frito-Lay's brand management decided that the unique new Mexican flavor variety would have greater growth potential as a separate, stand-alone brand. Thus, Frito-Lay launched the Mexican flavor as the Tostitos brand, with a new brand identity and new packaging graphics, thereby giving birth to one of Frito-Lay's most successful business ventures.

If a totally new brand or product line is to be introduced, the design assignment becomes even more complex. You are entering uncharted waters, and packaging will carry a heavy load of the communication responsibility. How can the package help in positioning the brand as new and exciting? How does the new brand differ from existing competitive brands? What benefits does the new brand offer over existing brands? How can package design convey the new brand's attributes? Where will the new brand be displayed at retail? How many facings will the new brand have at retail? Can the package design be uniquely different yet fit into the product category? These and many more questions suggest the broadest possible concept explorations—as broad as time and budget will allow.

Graphic Concept Explorations

Once the orientation meetings have been concluded, the graphic design process will most likely start with preliminary concept explorations to determine brand imagery and the best way to communicate product attributes. This may include explorations of logo styling, symbols, colors, typography, and a variety of visual treatments.

Take, for example, the introduction of Smilk, a line of fat-free milk available in six flavors, positioned as a nutritious, healthy, and good-tasting alternative to soft drinks. Although chocolate-flavored milk had previously been available, fruit-flavored, fat-free milk was a novelty. Extensive imagery explorations searched for solutions to several critical issues.

Problem: How to convince children thirteen years old or younger that Smilk is just as much fun and as delicious to drink as soda or juice and available in six popular flavors while at the same time convincing mothers that the product has essential nutritional value.

Solution (see Exhibit 7.1)**:** Gable-top Purepak containers suggest nutritional *white milk* to the mothers. For the children, graphics feature a bouncy, colorful logo to communicate fun; bright and dynamic background colors and unique flavor descriptors (Chuggin' Cherry, Grinnin' Grape, Raspberry Jazz, and so on) identify the flavors; a cow character named Smilkster appears on the front panels playing soccer, tennis, or basketball, adding a fun element to the side panels.

While the Smilk example describes package design for a new food product, the introduction of a new brand of cosmetics, a new golf ball, a new type of analgesic, a new toy, or a unique new garden tool will, of course, require totally different criteria and each will evolve into substantially different design approaches. The key to the successful introduction of a new retail product is to clearly establish how the new product augments existing ones and to develop packages that will communicate this in a compelling manner, attract attention, and intrigue the consumer to try the new product.

Courtesy of SMILK, Inc., by Cole Riggs Photography

Exhibit 7.1

Structural Concept Explorations

If the introduction of a new product, the relaunch of an existing one, or the addition of a product variety requires structural package design development, such as a new plastic bottle, a new paperboard carton, a new pouch, or a unique new opening or dispensing method, the same concerns as in graphic design development regarding equity, benefit, and category issues apply.

While the development of new packaging structures may require a substantially greater investment in time and money, it offers an extraordinary potential for long-range benefits. The successes of products marketed in unique proprietary structures attest to the fact that investments in such containers often pay off handsomely in profits. Coca-Cola's new plastic bottle resulted in a 40 percent increase in sales, Listerine's bottle redesign reestablished the brand's leadership position, and the inimitable egg-shaped container for L'Eggs hosiery, though now extinct, provided that brand with many years of unrivaled success.

To address structural design solutions, concept explorations start in the same way as graphic design, with a broad range of concepts. The difference is that these will be in the form of two-dimensional *sketches* representing a variety of structural concepts before explorations proceed to more costly three-dimensional structures. The sketches are often accompanied by written comments explaining particular features of the structural design concepts, or they may even be augmented by preliminary, rough three-dimensional mock-ups in the designer's effort to provide easier visualization of the structure.

Concept Selection

In most package design programs, concept development is punctuated by periodic meetings between the marketer's project team and that of the consultant for the purpose of evaluating the design alternatives and selecting a limited number of design solutions for refinement and further development.

It is generally advisable to select no more than three design concepts for further development. Selecting too many concepts will tempt compromising the decision process. By selecting a limited number of concepts from an initial broad range of design alternatives, the decision makers are forced to focus on the objectives for the package development and precisely define the relationship of the package designs to the strategic objectives. This especially applies when consumer feedback is part of the design development process.

How are graphic design concepts selected for the purpose of obtaining consumer feedback? Research in connection with the relaunch of the well-known line of Sudafed cough-and-cold remedies serves as a good example. From the initial design development of dozens of package design concepts, ranging from close-in designs (designs resembling the existing Sudafed packages) to several uniquely different designs, *two distinctly dissimilar* graphic concepts were recommended by the designers and agreed to by the marketer and the research firm. Each of the two directions fulfilled the criteria of maximizing shelf visibility, communicating efficaciousness, and improving product differentiation. Yet the differences between the two concepts were sharply drawn:

- One design direction related closely to the existing color-saturated packaging graphics, thus retaining their perceived visual equity among consumers (see Exhibit 7.2, top shelf).

- Another package design concept was primarily white, accentuating prescription-like efficacy while narrow color bands differentiated several product varieties (see Exhibit 7.2, bottom shelf).

Selecting these dissimilar design directions, each of which interpreted design criteria in a distinctly different manner, enabled the researcher to probe and identify specific consumer attitudes during the focus group sessions. Respondents, expressing a clear preference for the more efficacious graphics, helped Sudafed to gain a leadership position in the cough-and-cold category. Even when the Sudafed packages were recently updated, their efficacious look (white packages with color-coded bands) was kept intact.

While Sudafed's introduction of a substantially new look was risky and is not recommended for all brand repositioning solutions, the important point is this: Only by carefully selecting a limited number of distinctly different concepts for consumer feedback or judgmental selection, can management weigh the design options effectively and progress toward a final decision.

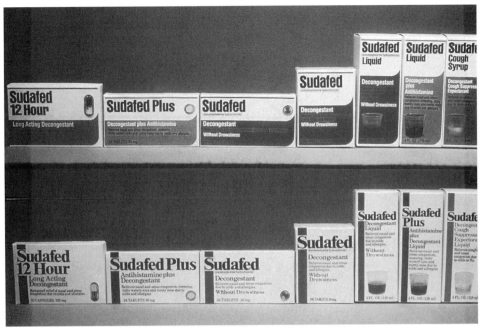

Courtesy of Warner-Lambert Company

Exhibit 7.2

Refinements and Modifications

Having selected the most suitable design concepts for further development, either through judgment or research, the designers will develop refinements and modifications leading to the final design direction. Although this phase is referred to as *refinements and modifications,* that description is somewhat simplistic and misleading. Refinements and modifications often require *substantial* additional design development to achieve the desired results. For example:

- If the selected concept utilizes a primary or corporate brand identity together with a subbrand logo (for example, Kellogg's Müeslix or Hefty OneZip), the relationship of the subbrand to the primary or corporate brand identity has to be carefully explored to achieve the right balance.

- If product pictures are part of the selected design concepts, a wide range of explorations may be required to select the most effective presentation of the product.

- If the product line consists of a variety of products, this may require extensive explorations to achieve the best possible product line segmentation or product differentiation system.

- Text, such as copy describing product attributes, that may not have been available at the start of the design explorations or that has been altered may lead to numerous adjustments of the design elements.

Thus, brand and product identification, product descriptors, colors, product presentations, and other key elements may require *extensive additional* explorations. It is not unusual, and it should not be surprising, that the refinement stage may be *lengthier* and *more complicated*—and yes, more *costly*—than the initial concept explorations.

This is sometimes difficult to comprehend by anyone not involved on a daily basis in the design development process, but it is perhaps easier to appreciate when one considers that the initial design phase was primarily concerned with generating *concepts*, while the refinement phase comes to grips with the nitty-gritty of specific marketing and design issues. It is all too common that the refinement stage is underestimated—timewise and costwise—because it is less exciting and stimulating than the concept development phase. In reality, this phase often ends up as the most meticulous, most time-consuming, and, therefore, most costly segment of the design development program.

Packaging Models and Mock-Ups

Similar to the graphic design development process, structural design development moves from the concept stage to refinements and modifications of the selected design concepts. In meetings between design consultant, marketing executives, internal technical personnel, and if possible, the designated package supplier, the structural concepts are thoroughly reviewed and a limited number (no more than three) mutually approved for further development. The designer is now able to move forward with refinements and three-dimensional mock-ups of the selected concepts, bringing the visualization of these a step closer to reality.

It is important that structural design for bottles or jars is reviewed three-dimensionally. Full-scale mock-ups made of wood, acrylic, plastic, or other appropriate material are provided either by the package designer, by packaging suppliers, or by model makers whose sculptural skills, materials knowledge, and special equipment can provide realistic looking three-dimensional representations of the packages. Aside from being essential to the packaging approval process, mock-ups often are required for consumer research or for preliminary stand-ins for advertising or sales promotion.

The mock-ups should be as close to the eventual production unit as possible and interpret all structural details. Depending on time and cost considerations, the materials used for the mock-ups may or may not be representative of the materials that will be used in actual production of the container. To visualize the appearance of bottles, for example, the mock-ups could be simulated by developing wood or solid acrylic models (see Exhibit 7.3). However, if the handling and dispensing of liquids are of primary concern, hollow, plastic mock-ups capable of holding liquids can be constructed to simulate the actual containers more realistically (see Exhibit 7.4).

Whenever possible, graphics should be applied to structural mock-ups and reviewed with the marketer's project team. This is essential to ensure that the selected packages live up to expectations and that all design elements are properly positioned on all sides of the packages (see Exhibit 7.5).

Occasionally, there is the temptation to bypass this step in an effort to save package development time and costs, especially if the package designs represent only evolutionary modifications. However, packages are three-dimensional objects that should *always* be viewed in three-dimensions before finalization. For example, the main panel of a round package, such as on a can or a bottle, represents only 40 percent of the total display area of the package. The remaining 60 percent flows around the sides and back of the circular surface. While the initial design explorations may be in the form of *flat* sketches for budgetary reasons or to expedite the design process, it is not enough to approve a design concept solely on the basis of preliminary renderings. Three-dimensional designs should never be approved for production until appropriate mock-ups have been constructed and all design elements accurately positioned on all sides of the packaging unit.

In that way, potential errors in judgment, such as brand name or product descriptors wrapping around too far on bottles or cans to be seen at one glance, or copy placed incorrectly on designated carton panels, can be caught and easily corrected.

Not only should bottles, cans, and cartons be mocked up, but packaging forms such as blister packs, pouches, and bags should also be viewed as total entities. Blister packs have flanges that over-

Exhibit 7.3

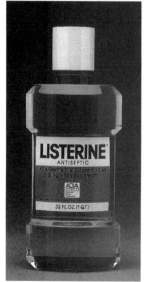

Exhibit 7.4

Exhibit 7.5

All courtesy of Warner-Lambert Company

lap and may interfere with legibility of important copy or affect the presentation of the products within the blisters. Flexible bags and pouches look substantially different when bulging with actual product inside. Large bags, such as those used for pet foods or fertilizers, are most often displayed horizontally at retail, so that the *end panels* become the primary panels.

Whatever the circumstances, the importance of developing three-dimensional mock-ups to visualize the final appearance of packages, whether graphic or structural, cannot be overemphasized. Consider the costs for developing three-dimensional packaging mock-ups as insurance against costly and time-consuming errors.

Package Design Finalization

After the final design direction has been selected by the marketer and approved for production, finalization of the package design is the last step in the package development process. It is as important and just as significant as the creative process.

The package finalization process does not enjoy the flexibility of design explorations. During the conceptual stage, it is relatively easy to make design changes such as reducing and enlarging design elements, adjusting copy, adding or deleting elements, changing colors, or even rearranging entire portions of the design.

The purpose of the finalization, as the term suggests, is to *finalize* the design process, so that packages can be produced and printed. There is less room for changes, and these always involve considerable expense and delays. Think of it as if you were building a house: it is easy to suggest changes when discussing the architect's renderings; it grows more complicated as you progress to the blueprint stage; it becomes downright prohibitive to make changes after the foundation has been poured.

There are, however, substantial differences between finalization of structural design development and graphic design development.

Structural Design Finalization

Structural design finalization requires final working drawings or blueprints that are critical for obtaining final approval from the marketer's project team and getting the package produced. Material specifications and dimensional details have to be 100 percent precise so that the package will hold, store, and protect the product exactly as planned. The slightest error in the package dimensions may affect the contents, structural rigidity, stacking strength, and weight of the package, or it may affect handling and filling procedures on the packaging line at the plant, not to mention the cost of correcting the package itself. Bottle and jar finishes have to be prepared accurately to make sure that the closures fit properly to protect and preserve the contents, avoid leakage, open and close easily, and prevent the product from spoiling during transport and storage. Paperboard packages have to be precisely prepared with regard to proportions, die-cuts, and print surfaces as well as machinability.

Graphic Design Finalization

Graphic design finalization requires similar exacting standards in preparation for print production. This involves many steps, including exact specification of the print area, bleed requirements, renderings of logos and various design elements, and preparation of final package copy. Typographic styling, size, and positioning, and the color of mandatory and legal copy must be specified. All copy must be proofread and approved by the marketer, often including the marketer's legal department.

Photography for Packaging

When photography is required by the package design, this can sometimes involve a complicated and time-consuming process. Many judgment calls require negotiations and compromises between the marketer, the designer, and the photographer:

- If the products are of a technical nature, interaction with appropriate technical personnel will be required.
- If food is involved, preparation will include searching for, finding, and selecting props, such as dishes, utensils, and background colors. The preparation of the food will be discussed with the food stylists and explored thoroughly. Should the food in the package, rice for example, be

shown by itself or on an attractive dish together with other appropriate food or condiments? Can small products, like cereal, be enlarged to show more detail without confusing the consumer as to the texture of the product?

- Does a product, such as a clock, look best facing front or as a three-quarters view? In what type of environment should it be photographed—office or home? What kind of props would best represent either location?

- If photographs show people or parts of people, such as faces, hair, or hands, the designer and photographer must review and select the most appropriate models for communicating the desired imagery. The same applies if photographs of animals are part of the graphic presentation on packages.

- If a child is shown on a package for children's medicine, should the child appear *in need of relief*, or should the child look happy as a *result* of having taken the medicine?

Packages, as compared to ad media, often remain the same for many years and are repeatedly reproduced. Models usually require residuals or buyouts for appearing on packages. This applies not only to custom photographs but also to the use of stock photographs on packaging. Financial negotiations with regard to depicting humans or animals on packages can often be prolonged and difficult to conclude. For that reason, it is advisable that contractual agreements regarding photographs of humans or animals on packaging be established well in advance of their use, even if only for preliminary concept development.

Contractual arrangements regarding residuals and buyouts of photographs are usually handled by the design consultant on behalf of his or her client. These should clearly spell out all obligations for payments by the marketer to the photographer, model, agent, or all of these. Particular attention is needed in specifying the *extent* of usage. Does the contract for using the model on packaging apply to a single SKU or a line of packages? Can the photograph be used during repeated press runs without requiring additional payments? What about showing the packages—which include the model's photograph—in TV commercials, print ads, and promotional media? Does the model or the model's agent require residuals for each of these uses? Can a one-time buyout provide for unlimited usage? If a buyout is agreed upon, may the model appear on packages of a competitive brand? If not, may the same model be used on packaging for noncompetitive products?

All of these issues require the designer's and the marketer's close attention if they want to avoid unexpected and annoying delays, potentially substantial expenses, disagreements, or even worse, legal action by the photographer, the model, or the agent.

Following the photography, there is another lengthy process: selecting and retouching the best transparencies or making arrangements with the color separator to achieve the necessary effects digitally. In some cases, several photographic components may have to be combined. Each step involves close coordination between photographer, designer, and color separator to achieve the desired effect.

For *hand-rendered illustrations or computer-generated artwork*, the same painstaking process applies during the finalization process. Illustrations are most often used when certain visual effects cannot be achieved through photographic means. For example, when product use or preparation instructions need to be communicated, hand-rendered illustrations or pictograms are often the clearest way to communicate this type of information, especially on packages that are marketed internationally, appealing to consumers who speak different languages.

After all of these details have been worked out and completed, when the artwork has been finalized, proofread, and approved by the client, and the disk containing the digital artwork has been forwarded to the color separator or supplier, there is no turning back. As the saying goes, The buck stops here.

Package Design Guidelines

Structural Design Guidelines

If *packaging structures* are involved, these require control of the physical packaging elements—such as materials, dimensions, weights, and manufacturing steps. Because these elements change frequently, resulting from production or economic considerations, they cannot be easily cataloged. They are, therefore, most often expressed in terms of technical *specifications,* or structural design guidelines, for internal purchasing, plant, and quality control personnel, as well as for package suppliers.

In most companies, these specifications consist of several integral documents. In *Fundamentals of Packaging Technology*, author Walter Soroka describes the specification process as follows: "A complete product specification is not a single document, but rather a group of documents describing all materials, components, and manufacturing steps that will result in the desirable product, as well as the characteristics that will define a quality product." For this reason, the technical guidelines for containers are usually documented and maintained separately by plant production personnel, engineers, or both.

Graphic Design Guidelines

If the brand represents a large number of SKUs—say, a hundred or more—the graphic design system will benefit from the development of a package design control manual (see Exhibit 7.6). The purpose of the package design control manual is to address the dos and don'ts of the package design system and provide guidelines to anyone involved in the design development process.

Graphic design control manuals not only will serve as a guide for package designers responsible for design conceptualization and modifications of packages but will also be useful for production personnel, package suppliers, and anyone else involved in the chain of package design implementation.

The graphics manual should include guidelines and specifications, such as

- logo styling and dimensions

- logo colors

- symbols and icons

- product descriptors and "sell copy"

- guidelines for typographic styling

- packaging colors (most often identified by the Pantone Color System® to which printers are able to match their printing ink colors most accurately)

- matched colors not available through the Pantone Color System®

- color adjustments for various printing processes

- photography and illustration styles

- guidelines for application of promotional bursts

- visual treatment of secondary panels

- design adjustments to accommodate certain production processes

- adjustments required for packaging substrates

- legal and mandatory specifications

- placement and dimensions of the universal product code (U.P.C.)

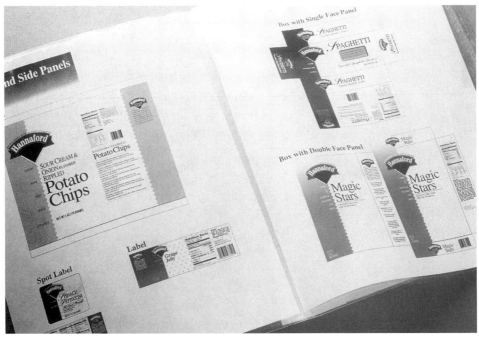

Courtesy of Hannaford Bros. Co.

Exhibit 7.6

Development of package design control manuals, like the design of the packages themselves, requires a step-by-step procedure like the following.

Step 1—Design Explorations

- design explorations of manual cover, pages, and section separators
- development of grids, diagrams, and package reproductions
- guidelines showing acceptable and unacceptable design applications
- development of copy clearly describing design dos and don'ts

Step 2—Modifications and Additional Pages

- modifications of page layouts and page contents
- additional page layouts showing a greater variety of packaging forms
- additional copy development

Step 3—Finalization

- final modifications based on approved design and copy
- development and approval of final copy
- development and approval of production artwork
- development of logo reproduction sheets (logos in a variety of sizes)
- development of color specifications
- specifications for manual reproduction

In order for the package design control manual to be accepted and implemented by all those who interact with package design and production within and outside of the company, it is important that the manual include the endorsement of the company's senior officer. A page addressed to all users of the manual, written and signed by the CEO, will identify the manual as official company policy.

Printed manuals require very special development and production skills. Responsibility for manual development includes recommending a printer known to be experienced in manual production and assembly, obtaining cost estimates for client approval, and supervising the production of the manual. In the absence of an internal design director or design staff, the responsibility for developing and implementing the package design control manual is best assigned to the consultant who developed the packaging program.

Electronic Package Design Control Manuals

Until recently, most package design control manuals, consisting of one to several volumes, have been issued in printed form. More recently, however, package design control manuals have been prepared digitally and made available to the users on disks (see Exhibit 7.7). Capitalizing on electronic technology has a number of advantages. Once the basic program for a digital manual program has been developed by a digital manual specialist, the digital information facilitates almost instant creation of new packaging units within the established design system. Style, size, and placement of copy are automatically assigned to designated positions on all sides of the packages and electronically adjusted to various packaging forms and dimensions. These digital capabilities also simplify the implementation of modifications, updating, and transmitting the manual information to company locations around the globe.

If developed electronically, the package design control manual must be implemented by individuals who understand package design and usage and are trained in processing the appropriate digital information.

Limited Edition Package Design Control Manuals

When only a small number of participants within a company, such as the company's internal design staff, are involved in the graphics

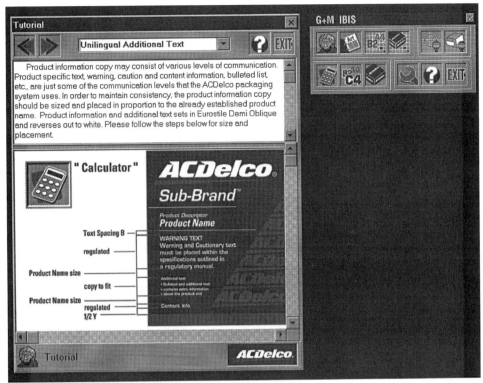

Courtesy of General Motors Corporation/ACDelco®

Exhibit 7.7

design process, the production of a printed, ring-bound volume or an electronically prepared control manual may not be economically practical. In that case, a *limited edition* manual would still be helpful in maintaining the design consistency of a multi-SKU brand line. This can be achieved by utilizing already existing artwork and digitally creating a few *hand-produced* copies to be distributed among those responsible for package design implementation.

Appointing a Manual Administrator

Regardless of whether the package design control manual is an extensive manual of several printed volumes or disks or a small one with only the most rudimentary design specifications, a *manual administrator* should be designated within the company. This can be the company's design director or someone in the marketing or advertising department. To prevent the manual from becoming outdated, procedures for incorporating manual changes and updates and providing assistance to manual users should be put into place. If the manual is an electronic one, an individual thoroughly trained in handling digital material should be responsible for keeping the manual current and disseminating manual information, changes, and additions needed by other company departments and offices.

8 Consumer Research: Navigating the Category

The use of consumer research in connection with package design has always been one of the most perplexing and controversial issues. The reasons for this are complex and produce as many proponents for as detractors from the value of package design research.

Some brand managers will not make *any* package design decision without confirmation through consumer research. Others believe in their own intuitive judgment so strongly that they rarely utilize consumer research for package design or ignore what they find. There are those who use research primarily to fortify their position when talking to top management and those who look to research as a replacement for making independent decisions.

Viewed objectively, these extreme positions show a lack of understanding about the purpose of consumer research, its potential and benefits, as well as its limitations. Knowing how to derive the best value from package design research can help make your package design decisions more reliable.

The brand management of many companies, as well as design consultants, utilize consumer research for a variety of packaging related purposes:

- to help develop design strategy
- to suggest possible design directions and modifications

- to confirm the appropriateness of selected designs as a support for marketing strategy
- to discover any negatives perceived by the consumer

Note that in listing these reasons for utilizing consumer research we used terms such as *to develop, to suggest, to confirm,* and *to discover.* Nowhere are we suggesting that the purpose of consumer research is to *impose* a decision on anyone. In fact, the first rule when considering consumer research for packaging is as follows: consumer research does not make decisions for you. But consumer research can furnish valuable information and knowledge that will *bolster* your ability to make intelligent judgments regarding package design.

Consumer research probes consumer behavior patterns, attempting to answer such questions as

- What are the shoppers' buying habits and shopping behavior in the category?
- What equity elements (for example, logo, visuals, color) do the current packages have? Do they have any?
- What are the strengths and weaknesses of the product(s)?
- What motivates consumers to buy or not to buy the product(s)?
- What motivates some consumers to prefer a competitive brand?
- How do consumers recognize and find the product(s)?
- In what way, if any, does the packaging influence consumers' choice?
- What are the communication or information priorities that guide consumers' buying decisions?
- What do competitive packages communicate about *their* products?
- What do consumers like or dislike about various packaging features in the category?
- Are the existing packages easy to transport, hold, dispense from, and close?
- What packaging improvements could influence consumers' choice?

Determining the Right Consumer Research for Package Design

Not too many years ago, marketing managers did not utilize consumer research extensively for package design development because they were either not sufficiently familiar with package-related research techniques—as opposed to those used for advertising research—or they felt that their intuition and marketing experience were sufficient to select the appropriate package design directions.

For many years, package designers were even less likely to utilize or recommend consumer research in connection with their package design development. Many of them were vehemently opposed to any type of consumer research that they felt would inhibit their creativity, thereby leading to conventional and *expected* package design solutions. Even today, some package design professionals voice doubts about the value of research and insist that their experience and creative capability are all that matters. These designers feel that subjecting their creative recommendations to mechanical and psychological testing techniques by researchers, who may be more oriented toward statistics than creativity, will inhibit originality and result in unexciting design solutions.

These concerns are not entirely unfounded. Many research firms that are commissioned by the marketer's research department to conduct package design research may be highly experienced in consumer research for advertising or other types of communications but less familiar with probing for the peculiarities and characteristics of *brand identity and package design*. There are substantial differences. In contrast to packaging, commercials and print ads need not compete for attention side by side with competitive commercials and print ads. Advertising tends to be temporal, changing frequently, while packaging must have staying power.

Elliot Young, president of Perception Research Services, a firm that has specialized in packaging-related research for more than twenty-five years, sums up the difference between consumer research for TV commercials and research for package design by stating, "With television advertising, the marketer is buying *time*, not space. In the store, the marketer is buying *space*, not time."

Packages, in contrast to the excitement that can be created in a TV commercial, have to achieve an entirely different set of objectives from advertising's:

- Packages are placed on shelves, hung on Peg-Boards, or are otherwise displayed at the point of sale—neatly or in disarray, depending on the attendance and efficiency of the retailer's personnel.

- Packages are displayed next to one or more directly competitive products and thus must be clearly distinguishable in the competitive arena.

- Packages must communicate their contents clearly and instantly, usually within a very restricted label area dictated by the container's proportions.

- Packages have to deal with a complicated array of technical requirements relating to production, printing, assembly, product protection, packaging equipment, shipping, and warehousing, all substantially different from the technical requirements of TV commercials and print ads.

For these reasons, a marketer who is considering package design research must understand the differences between ad research and package design research. Ad research probes primarily for recall, while packaging research probes for findability and quick communication. Recall, according to Elliot Young, is not as important in packaging as in advertising. The fact that a shopper recalls a product, while certainly of value, is less critical to the purchasing cycle than quick communication. Most important, according to Young, is the need for *justifying the brand's position* in the marketplace. Probing whether or not a package design has achieved this objective is where consumer research can be most effective.

Consumer research for packaging needs to probe an often complex set of issues relating to the proposed design concepts. This is especially critical when the subject of the research is the redesign of an existing package or packaging line—perhaps one that has been on the market for a long time and enjoys strong recall among shoppers but needs updating to respond to competitive pressures in the category or other marketing-related incentives. To *navigate the category*, research may need to probe for a variety of potential obstacles as well as opportunities in the context of

- shopping behavior in the category
- the consumer's decision-making process
- package design equities
- perceived values in comparison with competitive products
- priorities (for example, brand, variety, visuals, color, convenience features)
- perceived image based on package design
- aesthetic appeal

These and perhaps many more issues are critical to obtaining a better understanding of how to achieve package designs that will *justify your brand positioning.*

The resulting analysis will provide guidance regarding issues such as category shopping behavior, equity recall, value perception, strengths and limitations of the proposed design concepts, and purchase intent.

Determining the Best Consumer Research Procedure

The most frequently used brand identity and package design research methodologies are in connection with

- pre-design consumer research
- consumer feedback
- post-design consumer research

Pre-design consumer research, as the name implies, refers to research that takes place before actual design development has started. This type of research is especially valuable when no other marketing research exists for helping to establish design objectives or when such research is not sufficiently current to be relevant. It is also important to recognize that even when some type of marketing research regarding the product(s) or the product category is available, it may be mostly statistical and thus may not shed any meaningful light on *package design* issues.

When information that could be helpful in determining design criteria, such as shopping behavior and purchasing habits in the category, is not available, pre-design research can be useful not only to assist the design consultant to develop appropriate design solutions but to help you, the marketer, be in a better position to evaluate the design explorations. Package design without relevant information about the category is like flying a plane without a compass—it will most likely miss its destination.

Consumer feedback is research conducted at some point *during* the design development process. Its primary objective is to probe the appropriateness of various design concepts as they relate to the strategic objectives. Does design concept A support the strategic position of the brand as effectively as design concepts B or C? Which design concept communicates the marketing strategy most effectively and most quickly? Which design concept best involves the shopper? Which is easiest to find in the competitive environment? Which of several concepts should be retained, which should be deleted, and which provides the best opportunity for improvements through further development?

Using this information, marketer and design consultant will be able to evaluate intelligently which improvements of the design concepts might lead to the most appropriate solution—that is, what to change, what not to change, and whether or not there is a need for additional explorations.

Post-design consumer research is useful in *confirming* the final selection of a brand identity and package design program. This type of research could be in formats ranging from one-to-one interviews all the way to full-scale test markets. Many times, it may be sufficient to do a "disaster check" by means of a quick battery of consumer feedback that will confirm the effectiveness of the selected designs and relieve some of the nervous tension that is often part of the design decision process.

Research firms that *specialize* in package design research have developed productive research techniques geared to probing the appropriateness of package designs in relation to the marketing strategy and to assisting marketers and designers to make intelligent decisions.

Despite lingering objections to consumer research by a few professionals within the design community, most design consultancies have learned to appreciate the value of unbiased, consumer-driven feedback to their creative output. If the research is properly conducted, they no longer object to feedback out of fear of its inhibiting their creative output. In fact, a few design consulting firms have actually *added* research capabilities to their own services.

The involvement of design firms in research falls roughly into three categories:

- design firms that prefer to recommend outside professional research specialists who develop research methodology, conduct the research, and develop the research reports independent of the design firm

- design firms whose staff includes a research department with internal capabilities for handling consumer feedback in connection with the firm's design assignments and whose primary objective is to help the staff achieve precisely targeted design directions and to present their recommendations to the firm's clients with the benefit of consumer feedback

- design firms whose staff includes a research director who, while not directly responsible for research procedures, coordinates with clients' research departments in the selection of outside research firms, helps in developing appropriate methodologies for package design research, observes the research procedures, and reviews the results of the research firms' written reports

The choice of appropriate research firms and of research methodologies for package design research depends on the type of feedback needed to make design-related judgments. Such judgments must not be subject to the personal prejudices of individuals involved in the design development process, either on the marketer's or the designer's side.

There are many different research methodologies used for brand identity and package design research. The following describe those most often utilized in connection with package design.

Tachistoscope

Tachistoscope, often referred to as *T-scope,* is a technology that measures brand or product recognition and findability when individual packages or packages exposed in a competitive array are shown to the consumer for very brief intervals. Starting at a fraction of a second and gradually increasing in length, this testing method determines how quickly recognition of the brand or product will occur. Although the tests are usually combined with followup interviews of the respondents, T-scope tests emphasize the speed of consumer recognition of brands, products, and product information but often lack the ability to provide in-depth information about package design issues beyond impact and findability.

Eye-Tracking

Eye-Tracking, a related but much more sophisticated testing method, utilizes laser technology to trace the path of the eye as it scans a package or a shelf display, thus analyzing shopping behavior. This provides a diagram that shows the movement of the respondent's eyes as he or she scans the packages from design element to design element or from copy element to copy element on individual packages or from package to package in the competitive array (see Exhibit 8.1). This technique provides information to determine the shopper's *visual* priorities, that is, what the shopper sees first, second, and third, and to uncover packages and packaging elements that may be overlooked by the shopper. In addition, one-on-one followup interviews probe for issues that mechanical means cannot accomplish; for example, the aesthetic appeal of the package, the perceived uniqueness of the product, the ability of the packaging to stimulate product trial.

Focus Group Interviews

Focus group interviews involve controlled conversations with small groups of consumers—usually a representation of product users and nonusers in a specific category—by a moderator who is skilled in eliciting opinions and attitudes towards the category in general

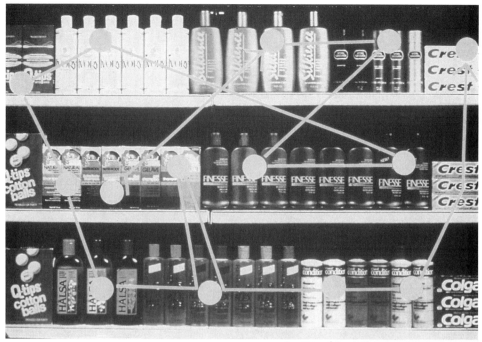

Courtesy of Perception Research Services, Inc.

Exhibit 8.1

and brand identity and packaging issues in particular. This method is popular among marketers and designers because it allows for a great deal of flexibility and is relatively cost effective. Its popularity results from the marketer's ability to observe consumer reactions and listen to their comments from an adjoining room behind a one-way mirror (see Exhibit 8.2). This provides the additional benefit of being able to evaluate consumer responses to the package designs even before the moderator's written report is available.

One-on-One Interviews

Similar to focus group interviews, the one-on-one interview method elicits from the consumer opinions and attitudes toward products, product categories, or package design directions through individual interviews with appropriate respondents, that is, users and nonusers of the products. These interviews can be conducted in or near the shopping environment, in the respondent's home or office, or at other locations, depending on the type of products involved and the type of respondents required. One-on-one

Exhibit 8.2

interviews can be particularly useful when the respondents are
individuals such as doctors or other professionals who have limited time for participating in testing procedures or respondents
whose relationship to certain products touch on intimate or sensitive issues, such as, in packaging of drugs for Alzheimer's suf-
ferers or AIDS patients.

Simulated Store Tests

A simulated store test requires simulation of an entire store section
or product category with the package design candidates displayed
among competitive or potentially competitive products. Respondents are asked to "shop" the section, and their shopping behavior
is observed and noted by the researchers. This technique also
includes interviews with the "shoppers" who are asked to comment
on what they have selected or not selected and why they made certain choices, with particular attention, of course, to the proposed
design concepts, which are included in the shopping display.

Full-Scale Test Markets

The most realistic consumer research technique is the utilization of full-scale test markets. This requires not only the production of a large number of packages filled with actual products to achieve realistic shopping situations but also the careful selection of the most appropriate test environment and the agreement by one or more retailers to use their stores as testing facilities. While this testing method is the most realistic, it requires a massive effort on the part of the marketer, the research firm, and the design consultant to organize and orchestrate the testing procedure. This method usually requires a lengthy period of time—several months to a year or more—to observe and monitor the movement of the test packages. Although this consumer research technique is the most ideal method of confirming the effectiveness of a new package design program, the cost connected with such test markets is, of course, considerably higher than those previously described. Despite the high cost, however, this may be the most productive method of package design research when it involves a major change for a well-known megabrand.

Dos and Don'ts of Package Design Testing

When a package redesign is consumer tested, the current packages should always be included in the testing. Results from testing the current packaging are used as a benchmark against which the respondents' reaction to the new packages can be compared. However, respondents should never be asked to compare the current package designs with the new designs, as this would never occur under actual shopping conditions. A cardinal rule for package design research is to select research techniques that will relate as closely as possible to actual shopping situations.

Respondents should never be encouraged to approve or disapprove package designs directly. Questions such as "How do you like this color?" or "Which design do you prefer?" should never be asked of the respondents, as such questions do not simulate realistic purchase situations, nor will they help in determining whether brand and product *communication* on the packages are on target.

Answers to such direct questioning are misleading, as they make the respondents, in effect, design directors, a qualification that is inappropriate and unrelated to actual purchase situations. Questions should always be directed toward the *product,* so that the respondents' comments will indicate whether the container communicates or does *not* communicate the desired product attributes and whether it supports the intended marketing strategy.

The effectiveness of directing attention to the *products* rather than the packages can be illustrated by the following example. Several years ago, marketers of a regional brand sold in the New England area tested design concepts by producing actual printed cans with three different graphic design concepts and filling the cans with their beer (see Exhibit 8.3). The testing took place among beer drinkers in markets where the brand was well-known. Respondents were asked to taste-test the beers in the three cans and describe their opinions about the differences between them to the researchers. All cans contained the *same* beer, yet the descriptions of the beer expressed by the respondents varied substantially, ranging from "light" to "heavy" and from "watery" to "full bodied."

Exhibit 8.3

Since the beer in the three cans was exactly the same, the responses showed conclusively that the packaging *graphics* influenced the beer drinkers' perceptions of the products. This was exactly the information the marketer was looking for. The winning design was eventually produced and successfully marketed, based on the comments expressed during the research.

Consumer research can and should be an important segment in the package development process. While research may not guarantee conclusive decisions, it can be extremely helpful in focusing on design objectives and their appropriateness for a specific marketing strategy. While consumer research does not necessarily determine the final design direction, consumer research can and usually will lead to specific recommendations.

At the very least, packaging research will confirm the appropriateness or inappropriateness of the verbal and visual communication elements of the designs. Research can also be effective in separating personal partiality from consumer preferences. For example, when the design consultants to Breyers ice cream recommended black backgrounds for their packages to highlight the appetizing product photography, Breyers's management judged this as too radical in a category steeped in the tradition of white packaging for dairy products. However, packaging research conducted independently confirmed the designer's recommendations, and the Breyers brand in the new packaging subsequently became the leading nationally sold ice cream in the United States.

Another example involves baby products and pet foods. When packages for such products feature pictures of babies and animals, respectively, to appeal to the empathy of the purchaser, the selection of pictures usually becomes a highly personal and subjective matter. For this reason, package designs containing such visuals will benefit from research to obtain consumer feedback that is not tainted by the marketer's—or the designer's—personal biases.

The research procedures described above also apply to structural packaging development and often lead to refinements and modifications. The shape of a bottle, the placement of a handle, the ease of opening and closing, the size of the container, and the tactile feel of the packaging surface or shape will all contribute to acceptability or nonacceptability by the consumer.

One aspect of consumer research that has not been discussed so far is the ever present possibility that package designs being tested will be totally rejected by the respondents as being unsuitable for the products. While this is not a pleasing thought for either the marketer or the designer and while it does not occur frequently, there is always a possibility that the entire team's judgment has been channeled in the wrong direction.

When the marketer of a major diet food line became aware that consumers regarded the brand's products as bland tasting, the marketer attempted to reverse this perception through packaging graphics showing sumptuously prepared food on elegant dishes and table arrangements.

The marketer, presenting these magnificent-looking packages to focus groups, discovered that this approach was totally wrong. It did not improve consumer perception. On the contrary, consumers indicated that dietary food "just couldn't look *that* good." They rejected the packaging concepts as inappropriate and misleading.

No matter how difficult it is for marketing managers and the design consultant to accept such a reversal, it is obviously better to find this out at the consumer research stage than to market products in packages that will prove to be a costly disaster in the marketplace.

Consumer research can prevent you from introducing products in packages based entirely on intuitive decisions and mitigate the risk of potential failure. Most important, since total failure of a design program often indicates problems beneath the surface of the marketing strategy, consumer research can be helpful in reevaluating marketing strategy and steering the strategy and subsequently the package design criteria in directions that will be a better fit for the brand.

Remember, consumer research never *imposes* final decisions on you. *You* are always in charge. *Your* judgment will determine the ultimate direction that the brand will take. While consumer research is a valuable tool to *navigate the category*, the ultimate choice of which direction your brand and its packaging will follow will always be based on *your* vision of what your brand's long-range marketing strategy should be.

9 Store Brand Packaging: Friend or Foe?

William Shakespeare asked, "What's in a name? That which we call a rose by any other name would smell as sweet." Call them private labels, store brands, house brands, exclusive brands, own labels, or any other name, it is important to understand that while the descriptive names may not matter, the proliferation of store branded products is becoming increasingly significant and demands a new way of thinking about the marketing of retail products.

While the private labels of supermarkets, discount outlets, clothing stores, and a variety of other retail chains were once considered second-best and sought primarily by bargain hunters, they are now catering to consumers who are more discriminating, seeking quality products *along* with reduced prices. And as mega-retailers gobble up smaller chains, their increasing control over what they sell, where they sell, and how they sell adds significance to what their packaging looks like. Along with improved product quality of store brands, the shelf appeal of store brand packaging becomes as meaningful as their pricing in enticing consumers to purchase the store brands and to return to the brand after their initial purchases.

Retailers are beginning to understand that *all* products available in their stores are *brands*. It matters little whether these are category specific brands marketed by manufacturers or marketed as the retailer's *store brands* that cover a wide and diversified range of products. The important thing to understand is that the shoppers' perception of the difference between a manufacturer's brand and a store brand has been constantly narrowing. Thus, it becomes increasingly critical for both the manufacturer and the retailer to clearly define and communicate their philosophy and price/quality positioning to the consumer.

Although manufacturers cry foul, the fact that cannot be ignored is that the retailer is fighting for survival just as passionately as the manufacturer. But the retailer has a different perspective than the manufacturer: to make a meaningful and lasting impression and entice shoppers to return to the stores, the store brand must provide a desirable alternative to other brands for reasons of quality, not price alone. "The biggest cliché about private labels," says Marty Gardner, store manager of Wegman's, a highly successful supermarket chain in upstate New York, "is that they are just as good as any national brands but can be offered at a little less. We say, make them one of the top reasons for people to shop at our stores."

How Store Brands Got Started

In his book *Private Label Marketing in the 1990s,* Philip Fitzell describes how private labels in the United States started by using mail-order houses and gradually evolved into retail businesses primarily through the auspices of A&P. In the 1960s, private labels amounted to an impressive 35 percent of A&P's total store volume. Antitrust investigations trimmed this back in the 1970s to about 15 percent, but it has since rebounded, supported by two strong A&P label programs, America's Choice and the upscale Master Choice (see Exhibits 9.1 and 9.2).

In 1926, twenty-five retailers affiliated to form the Independent Grocers Alliance (IGA). By 1930 IGA labels appeared in ten thousand food stores and were considered another national brand. Their example was quickly followed by Kroger, JCPenney, and Sears Roebuck, whose growth and visibility in marketing power were augmented with the help of television. However,

Courtesy of The Great Atlantic & Pacific Tea Company

Exhibit 9.1

Courtesy of the Great Atlantic & Pacific Tea Company

Exhibit 9.2

these programs quickly degenerated into a multitude of independently branded product lines, thus missing the opportunity to build strong memorability and brand equity for the products available in these stores.

This situation was further complicated by the appearance in the 1970s of generic "no-name" labels, a concept that started in France and Germany. While popular among bargain seekers, the fallout from the generally poor quality of the no-name products at that time further depreciated the perception of store brands.

This unfortunate trend was reversed when David Nichols, then president of Loblaws Stores in Canada, developed the idea of an upscale store brand, President's Choice. The basic idea behind President's Choice was to attract consumers to Loblaws supermarkets with merchandise available only in these stores and to raise the level of perception of the stores themselves. The eminent success of this program is reflected in the fact that President's Choice brand products are now available not only at Loblaws in Canada, but in other store chains throughout the United States and even overseas.

The potential of successfully marketing store brands was particularly recognized by European retail chains as European nations recuperated from the ravages of World War II and rebuilt their economies. In fact, retailers such as Sainsbury in England, Carrefour in France, and Aldi in Germany have gone far beyond the store brand strategies of U.S. retail stores. The United Kingdom's largest food retailer, Sainsbury, has a 19 percent share of trade in the United Kingdom and generates 54 percent of its sales from its private labels.

This degree of control of store inventory did not come all by itself. The European store label programs span a wide range of products and categories with enormous commitment to quality and variety surpassing what many European brand name manufacturers provide. European retailers devote a large pool of experienced marketing management to their private label programs because they consider the private label a powerful marketing tool as well as a significant contributor to bottom-line profits. It's surely no accident that in the United Kingdom private labels are referred to as *own labels*, signifying the stores' deep commitment and pride of ownership.

Do private labels in the United States get that degree of attention? It still seems to be difficult for many retailers to accept that as the dynamic of their business is changing, if they wish to convey the store's quality and commitment to build consumer loyalty, their private label programs must change. This is not likely to happen with retailers who still rely on their purchasing departments to handle package design assignments or delegate this responsibility to independently operating category managers, as is the case with certain mass merchandisers.

Without an overall store brand philosophy that has the approval of the highest level of corporate leadership and experienced implementors to develop and implement the program, store brand package design will be fractionated into many disconnected parts that fail to build equity and customer loyalty.

Building a Store Brand Packaging Program

While manufacturers' brands are most concerned with issues such as category competition, store location, benefit communication, and advertising, store brands cover such a wide variety of products that they must sacrifice some of these specific concerns for the good of the total image of the store brand and the store environment. For this reason alone, designing a store brand must be approached in a way that is substantially different from designing brand identity and packaging for a manufacturer's brand.

Store brand package design programs are a long-range, franchise-building program, reflecting the stores' total image and determining how the retailer wants the consumer to feel about their stores. For long-term profits and success, retailers must handle their store brands not just as labels that identify the products but as part of the total identity program that will enhance the stores' image in the mind of the shoppers.

Despite these differences between the packaging concerns of store brands and those of manufacturers, retailers who initiate a store brand program or update an existing one must take the same preliminary steps as any marketer to analyze the current status and future needs of the brand prior to the launching of a brand identity and package design program.

Like the analyses that precede design programs for manufacturers' brands, retailer management must address a whole
range of strategic issues that will impact on the direction the
design program will take.

- What is the retailer's image among current shoppers?
  What does the store chain's name communicate to the
  consumer?

- Why do the consumers shop in the stores? Is it
  convenience, price, selection, quality, or friendly service?

- What are the demographics and pyschographics of the
  shoppers? Does research about the shoppers exist, or is
  qualitative feedback from the consumers needed prior to
  store brand development?

- Do the current store identity, packages, and
  promotional material accurately reflect its philosophy
  and price point?

- How should the store brand be marketed? Should it be
  part of an umbrella program or be positioned as the
  counterpart of specific brands in various categories?

- Does the store brand serve the current consumer base as
  well as it should?

- Can the consumer base be expanded with additional
  store brand lines?

- Should the brand accommodate one or more tiers
  (economy and upscale)? If the choice includes an upscale
  tier, should it cover a broad range of products or just
  certain products that lend themselves to the perception
  of high quality?

- Should boutique products, such as bakery goods, fall
  under the stores' brand umbrella look, have their own
  unique identities, or be subbranded?

- How will the store brand relate to other media—store
  signage, merchandising displays, posters, store personnel
  uniforms, and so forth—to facilitate linkage of retail
  identity and store environment?

- What current procedure is in place for initiating and approving brand identity and package design requirements? Who will ultimately be responsible for implementing and maintaining the brand identity program?

Only when these and other issues have been resolved and an overall brand architecture has been developed and approved by senior management can the design program begin. There are many steps along the road to developing a store brand design program that are normal procedure with manufacturers' brands but are not always common with retailers. Consumer research, for example, which can ensure that the design program meets its original objectives, is rarely applied to store brand package design. Coordination between retailer personnel responsible for package design and those handling such items as store signage, shopping carts, uniforms, and rolling stock is the exception rather than the rule. Even coordination between the retailers' product categories is unusual. Resolving these issues at the beginning of a store brand design program is as important to the successful outcome of the program as the design program itself.

Optimizing the Effectiveness of a Store Brand Program

How then can retailers optimize the effectiveness of a brand identity program? How can the retailer build brand equity and store loyalty through the medium of packaging? There's no single answer to this, but the following rules may contribute to a recipe for success.

Rule 1: A store brand must function as a brand that sustains the store's image. "Store brands," says Marty Gardner, brand manager of Wegman's, "are far more than margin boosters. They are a way of reinforcing the store's identity as a shopping destination." By treating store brand programs as long-range franchise-building programs rather than just label design, the quality of the merchandise will be communicated effectively while building customer recognition that will create a solid base of consumer loyalty for the future. This means that those responsible must look at store design, store signage, posters, sales promotion, uniforms, and, of course, packaging (in short, everything that the shopper sees when entering the store environment) as icons that will add to leading the shopper to a positive perception of the store and its brand.

Rule 2: Define the store's point of difference. Retailers must define the store's point of difference from competitive retailers. They must clearly communicate their philosophy to their target customers with packaging that reflects the true quality of their products. This could mean many things: stores that target economy-minded shoppers should address this consumer segment with packaging that will neither look cheap nor imply luxury; stores that market fashions for men and women should capitalize on highly styled brand identity; stores that sell electronic products may want to look high-tech; stores catering to expectant mothers need to reflect the anticipation of motherhood. Packaging must differentiate and relate to such image objectives.

Rule 3: Reflect the price/value strategy of the store. There is always the temptation to try to make the packages and the products displayed on them as tempting as possible. But this can backfire, especially in the case of supermarkets and mass merchandisers who cater to very conservative, budget-conscious shoppers. Care must be taken that package appearance does not oversell the product. At the same time, cost-conscious customers who find their neighborhood stores poorly maintained and store brand packages generic-looking may perceive this as being condescending to them and will seek alternative stores whose appearance and packaging communicate that these retailers appreciate their patronage.

Rule 4: Be unique; you can't gain equity by cloning leading brands. "The brand's best longtime defense is to invest in innovation to keep their edge and to continue to build brand equity," says *Brandweek* magazine. A brand's packaging is loaded with psychological implications. The consumer visualizes the product through the package's shape, colors, symbols, and words and forms an opinion about value and performance.

When store brands copy and try to look like the leading national brands, such as the store brands of many discount drug chains and certain mass merchandisers, this will only breed confusion among consumers and in the long run fail to build equity in the store brand. Imitation may be a form of flattery, but cloning a leading brand to mislead the consumer into picking up the clone by mistake (and who can claim that this has never happened?) is simply deceitful. If a store brand is to be successful, it must exude confidence in its own distinct personality. This strategy requires courage and commitment on the part of retailers but can pay off handsomely in the long run.

Rule 5: Design the retailer brand's packages to reflect the retailer's commitment.
Don't develop a rubber-stamp program that becomes so consistent that it provides no design flexibility. While consistency of product quality is a desirable characteristic, appearing to be *overly* consistent can backfire if it's interpreted by consumers as representing a lack of caring and commitment. The dairy category is different from the cereal category and from paper products. Such categories should not match each other. Designing store brands that relate to various categories while maintaining a consistent overall brand identity is the key to good store brand packaging. Many excellent store brand programs have been developed in which retailers are able to compete with other brands in certain categories, yet maintain a sufficiently consistent overall brand look.

Rule 6: Renew excitement with each new product introduction.
When introducing a new store brand program or repositioning an existing one, retailers have the advantage of being able to control and maximize this event across all available media, including advertising, in-store promotions, shelf positioning, end caps, trailers, and many others. They have the unique ability to use every inch of their stores to promote the brand and use all of these components as part of the overall marketing strategy to communicate the value and benefit of the new or improved brand identity and packaging program. This opportunity should not be lost in the course of developing the design program.

Rule 7: Monitor your store label program constantly. Nothing runs by itself; your program needs constant monitoring. Just as your car needs occasional checkups and you need periodic physicals, store brand packaging needs to be updated continuously. It's not enough to develop a store brand program and assume that once launched, it will keep moving in the right direction. Consumer research and sophisticated scanner technology can probe for the continued appropriateness of packages in their categories and their suitability in relation to changing lifestyles, shopping habits, and preferences.

Equally important for maintaining store brand programs that cover thousands of SKUs is the need for establishing a procedure for initiating, coordinating, approving, and implementing packaging and other design components that relate to the store brand identity. This applies to the design program introduction as it does for items that will be implemented after the introduction. To keep the program on track, it is best if an individual at the executive level spearheads and coordinates the maintenance of the program.

Rule 8: Top management must be fully committed to support the brand strategy. Make no mistake about it—without the full support of top management, the store brand program will be condemned to failure. At Loblaws, David Nichols took personal charge of the President's Choice program when it was being developed, and thus it became one of the classic examples of how to create a successful store brand program. Tom Stevens, then a vice president of Loblaws, confirmed this commitment, explaining that "if retailers are going to develop private label programs, they must develop *marketing* skills and should turn to *marketing* executives to handle their own private label marketing."

Not All Store Brand Programs Are Alike

Because of the differences between the shopping environments of various types of stores, ranging from the traditional supermarkets to the sprawling landscape of wholesale outlets to the sophistication of fashion centers, each must approach a brand development program in a distinctly different manner.

This can be exemplified by the following two sharply contrasting case histories. One is a branding program for Hannaford, a chain of supermarkets serving mostly lower income shoppers, and the other is Ann Taylor, a clothing store chain designed to attract fashion-conscious professional working women.

Hannaford, headquartered in the northeastern United States, operates about 150 supermarkets in that area. Most, but not all, carried the store name Shop 'n Save. As the company expanded into new territories by acquiring other store chains, consumer research pointed out several important issues that reflected on the stores' overall image: the store name Shop 'n Save conveyed low price and low quality; shoppers considered the Shop 'n Save packages generic-looking; and the excellent quality of their products was seriously understated.

This information convinced management of the need for a number of dramatic changes. They called on their brand identity and design consultants for assistance and effected the following changes.

- New stores were named Hannaford based on the high regard for the parent company in its home territory (see Exhibit 9.3).

- The store brand packages were completely redesigned, using the Hannaford name as brand identification (see Exhibit 9.4). A new brand architecture divided the brand line into three segments (food, pharmaceuticals, and household products) and specified type styles, background patterns, and colors for all segments of the new packaging program.

- To further add to the impact of the new identification, the program was expanded to include new external and internal signage, colorful new decals for identification on trucks and trailers, and attractive uniforms for store personnel and drivers.

- A graphics manual developed by the design consultants enabled an economical procedure for implementing the design program to more than two thousand SKUs by the design department of the

Courtesy of Hannaford Bros. Co.

Exhibit 9.3

Courtesy of Hannaford Bros. Co.

Exhibit 9.4

company's category-specific broker without the risk of detouring from the original design concept.

Thus, by carefully analyzing its options, this company was able to revitalize its image of an expanding store chain, creating excitement among shoppers and resulting in sales increases across all stores.

In contrast to the already-described supermarket chain, Ann Taylor stores specialize in fashions for professional working women. This company sought to develop and extend its brand image by creating a retail environment that reflects the lifestyle of Ann Taylor customers: women who combine confidence with curiosity and femininity with adventure.

Their design consultants created a sophisticated, award-winning identity program that reflects the elegance of Ann Taylor's townhouse headquarters on New York's Madison Avenue. The program includes the following features:

- Store packaging uses sophisticated, classic type styles, natural materials, and decorative graphic accents (see Exhibit 9.5).

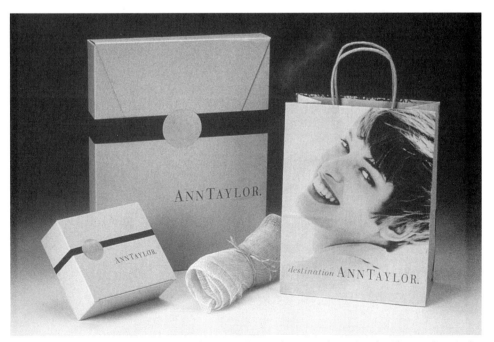

Design firm: Desgrippes Gobé & Associates, Creative Director: Peter Levine

Exhibit 9.5

- Packaging for their fragrance bath and body line, Destination, reinforces the uniqueness of the brand identity through the use of such packaging material as gunmetal, glass, wood, and textured papers.
- A separate and similarly distinctive brand identity and packaging program for Ann Taylor Loft, an Ann Taylor division, offers off-price goods.

The goal of brand identity and packaging for retail stores differs from that of manufacturers in a very distinct way. Design criteria for manufacturers' packaging address the primary objective of communicating the specific benefits of each of their brands and products. Retail identity programs and store brand packaging, on the other hand, are geared to the need to achieve the overriding goal of creating a shopping environment in which shoppers feel comfortable. The ultimate objective of store brands—not always understood by marketers—is not simply to create a devious scheme for competing with manufacturers' brands but to fulfill the retailer's need for securing the loyalty of the shopping community and expanding the customer base by attracting new shoppers.

If the retailer thus succeeds in attracting more shoppers, this not only contributes to increased profits for the retailer but creates, perhaps paradoxically, additional exposure of *all* brands in the retailers' stores, *including* the manufacturers' brands.

10 Package Design for Special Markets

Packaging of consumer goods in economically developed countries is one of the phenomena of our modern era. Who could visualize today's marketing environment without packaging to hold, protect, and enhance the products offered for sale? Many packages target a broad and largely undefined audience, such as, for example, basic ice cream flavors, batteries, or toothpaste. Even these product categories are often subdivided so that certain portions of the brand lines are targeted to meet the needs of specific consumer lifestyles or tastes. Still other categories—such as cosmetics, alcoholic beverages, fashion accessories, hardware, or automotive aftermarket products, to name a few—target consumer segments with specific behavioral needs with packages that address each of these in different functional and visual ways.

While each of these categories merit separate discussion, there are two important consumer segments that so distinctly differentiate themselves from general consumer profiles that they need to be addressed in greater detail: the very young, ages one through about ten, and the mature consumer, ages fifty and over. Pointing toward opposite ends of our life span, these two consumer segments are so vast and diverse that packaging for them deserves closer scrutiny.

Packaging for Young Consumers

If you have ever watched a youngster unwrap a gift that came in a package, be it a game, a doll, a construction kit, or a toy drum, you will be familiar with the sight of little hands frantically dismantling the package and disgorging the contents. But watch what happens next! After playing with the toy, the child will carefully return it to the package—albeit often with some parental prodding—which continues to be used as a storage unit and a means of identification among other toys.

This then is the first distinguishing feature of packaging for the young consumer: the package serves not only as a selling tool for the product by means of a distinguished structure and colorful graphics designed to attract and excite both buyer (parents) and receiver (kids) but also as a repository and, occasionally, as an integral component of the product.

Targeting Kids

Package design for products that appeal particularly to kids requires a totally different mindset by both marketers and designers compared to packaging for the adult consumer. Yet the marketing of such products has to be approached in the same businesslike manner of setting strategic goals, just like packaging for any other industry. The size of the kids market is enormous, estimated by various sources to range between $130 billion to $150 billion. This is thanks to the medium of television, which, more than any other source, contributes to the recognition of brands and products by kids, starting at a very early age.

Brand recognition, an important goal in package design for any product, assumes an even more critical position in marketing products that are meant to appeal to children. While the adult shopper can be swayed into picking up a product other than the one he or she had originally planned on when entering a store, there is no such flexibility in the child's mind; try to substitute a similar-looking doll for Barbie or an unknown brand for Power Ranger Turbo figures, and you will run straight into a brick wall.

The fact is that one of the first things a child learns is *brand recognition*. At age two, children are starting to recognize brands by name. The brand then becomes an object of their desire, and as they grow older, they become more and more specific as to their choices.

While this may appear simple enough, changes in society weigh heavily on the development and behavior of young children. One out of four American children now grow up in single-parent households. Seventy percent of mothers work full-time, requiring kids at ages six through twelve to do some of the home chores, including shopping and occasionally cooking their own food. Compared to twenty or thirty years ago, such family situations require kids to become self-sufficient at an earlier age and adjust to adult lifestyles more expediently than their intellectual growth may tolerate. For that reason, outside influences, such as television and other types of communications including packaging, have a particularly strong effect on the vulnerability of children during their growth years. Marketers understand that this can be both a positive and a negative element of marketing to kids. Kids will accept brands and products that fire their imagination and that they consider "cool," but they can also be cynical and will reject a product more quickly than adults when they feel that an ad or a package overpromised or was misleading.

But the greatest effect on marketing to kids is probably the influence of peer groups, especially when it concerns clothes, toys, and food. Some marketers have had huge success by taking advantage of the peer influence phenomenon. Kids discuss their favorite products among themselves and identify with various peer groups who favor highly promoted brands such as Sony, Disney, Nike, The Gap, and Kellogg's. The brand identification characteristics of these products—be it the bouncy Disney script, the Nike swath, or the Kellogg's logo—are quickly absorbed by kids at an early age. When they are not shoppers on their own, they will lobby their parents to buy the brands and products they favor (see Exhibit 10.1).

Research conducted by the Yankelovich organization in 1989 identified some of the product categories where purchases by adults are heavily influenced by pressure from children:

Courtesy of General Mills; Johnson & Johnson Consumer Products, Co., a
division of Johnson & Johnson Consumer Companies, Inc.; and Kraft Foods

Exhibit 10.1

80 percent of children influence purchases of breakfast cereals.

61 percent of children influence purchases of toys.

60 percent of children influence purchases of soft drinks.

28 percent of children influence purchases of toothpaste.

Brand recognition and the appearance of the brand's packaging in these and other categeories undoutedly play a major role in the children's decision-making process.

There Is No Single Kids Market

To make things even more complicated for the marketer, the kids market is more fractionated than almost any other, especially considering that it covers a life span of no more than about ten years. To begin with there are, of course, substantial differences between what interests girls and what interests boys. According to research conducted by CMS Kidtrends Reports, girls are generally more nurturing, creative, and expressive, while boys are more physical and competitive and enjoy outward activities. When designing packages for the kids market, these differences must be taken into serious consideration.

In his book *Kids as Customers: A Handbook of Marketing to Children,* James U. McNeil describes the influence of kids on three distinct markets and several demographic segments. The three markets are as follows:

- the primary market—children who have needs and wants and the authority and willingness to spend their own money to fulfill those needs and wants

- the influence market—children who directly and indirectly influence household purchases, including the acquisition of major products by their parents

- the future market—children who will become the consumer of *all* products as they grow older

McNeil further subdivides the kids market into several segments that influence not only the type of products targeted specifically to children but also the packages designed for these products:

- demographic segment—young, middle-aged, and older boys and girls

- lifestyle segment—children influenced by the lifestyles of parents, such as intellectual, outdoorsy, and so on

- benefit segment—users of low-cost, educational, or performance products

- usage-rate segment—light or heavy users of certain products

It is clear that marketing to such a market—one that is not only fractionated but compressed into the narrow time frame of childhood—is no simple matter. Marketers must be sure to have a precise profile of the young audience they address. This requires the same detailed attention as products for any other market, including clear strategic objectives based on category understanding, sensitivity, and good consumer research.

Creating Packages for the Kids Market

If, as noted earlier, shopping by kids starts at the age of six or seven, what should the packages for them look like? How should they be branded? What should they say about the products inside? What's the most effective way to communicate to children what's inside the package? What will get them excited about the product when they see the package?

Graphics—To be successful in today's global market, manufacturers of toys and other products targeted at kids are continuously exploring innovative products and packaging concepts that will gain the loyalty of an increasingly sophisticated young audience. The research firm Packaged Facts cites a 35 percent growth in sales of toys targeted at newborns to five-year-olds in the United States between 1991 and 1995. Products aimed at this age group cover a wide range, from rattles to electronic learning aids. To fulfill their need, the rising number of two-income families and a larger population interested in multicultural products create a growing pool of parents who encourage the developmental needs and capabilities of their children. This creates an ever growing array of "edutainment" products—products that link playthings with the learning process of kids—making the ability to communicate the attributes of these products more and more complex.

Buyers at mega-retailers, such as Toys "R" Us, have an ever increasing input on products and packaging at the early design development stage, and many of the best store promotions combine fun and learning by creating customer-friendly environments that combine the romance and magic of toys with interactive teaching. There is an increase of "hands-on" packaging in these stores that encourages children, as well as adults, to try the toy at the point of purchase. Packages for the kids market should be designed to accommodate this.

At the same time, syndicated television and film programs lead to more and more *licensed* products. Packaging graphics for toys and food, such as certain cereals and beverages, feature the personalities who appear in these programs to lure the kids who watch and adore them. By recreating the personalities' actions through colorful illustrations that amplify their activities, the packages will create a "have to have it" selling climate to attract their young audience. A toy's success often depends on whether its package can create a link with the child's imagination and accommodate the child's natural play instinct.

However, not all packages for the kids market require action-packed graphics. For some products, the ability to see and touch the actual product is more important, suggesting open windows or other packaging structures to enable this. Products such as dolls or toy trucks fall into this category. Illustrations or photographs will not achieve the realism that children require to help in their decision-making process. Kids are very realistic and interactive; they like to find out what certain products look like, feel like, and how they work before they devote any interest to them.

Names—If we agree with the statement expressed earlier in this chapter, that kids start recognizing and memorizing brands as early as two years old, it should be clear that the brand name or the name of the specific product must stir their imagination. Intriguing names help children remember the products. Names are of a few different types:

- *descriptive names* that hint at the appearance or function of the products, such as Play-Doh, Super Soaker, or Mootown Snackers

- *nonsensical names*, such as Frisbee or Nintendo

- *licensed names* relating to TV programs or movies, such as Star Trek or Teenage Mutant Ninja Turtles

Another way of creating brand names that have meaning to kids is through extending a line of adult products and using *derivative names,* such as Mini Oreos, Crest for Kids, Sports Walkman, Levi's for Kids, Johnson & Johnson Poos, and Band-Aid Hot Colors and Glow in the Dark. These products appear in packages that relate to those used by their parents but are jazzed up through more catchy names and colorful graphics that appeal to kids.

Copy—Copy on packaging for the kids market depends on what type of product is contained in the package. Kids expect marketers to be truthful about products and examine packaging copy carefully. In their book *Marketing to and Through Kids*, Selina S. Guber and Jon Berry describe it this way: "Before deciding how much to put on the package, you should learn from your consumers (that is, kids) how they will be using the product and how much time the package will stay around. The longer it will stay around, the more prominent a place it will have in the household. And the more time the consumer will spend with the package, the more information it can hold."

The authors go on to explain that some packages that stay around the house a long time, such as cereals, intrigue children who like to look at the stories, puzzles, and games that appear on the secondary panels of cereal packages while their parents are busy reading the morning newspaper.

Packaging of many children's products that need instructions on how to use or assemble them should include as much information as possible in language understandable to kids of the age level to which the product applies. Kids want to know everything about the product: how it works, how to put it together, what to do with it, and what the package includes or does not include—such as batteries. If you have ever watched kids assemble a toy or prepare to play with a new game, you will know how intently they read the instructions.

But not all packages have the same needs. Candy wrappers, for instance, need virtually no copy other than brand identification, as they are discarded as soon as the product is purchased. Hence, most candy manufacturers opt for making the brand logo the most prominent copy on candy wrappers. These take up virtually the entire main display panel, its main object being to create display impact and memorability.

Color—Perhaps no other element of packaging for products targeted at kids, especially toys and certain food, is as critical as color. Young children are attracted to the vibration of bright packages, and today's packages for products for preschoolers especially reflect this. There was a time, twenty or thirty years ago, when toy packaging was printed in two flexographic colors, and designers could select these—so long as they were pink or blue. Today, of course, with full-color printing an accepted method for virtually all packaging, this is no longer a restriction. Nevertheless, the tendency to use pinks and blues in packaging that specifically targets boys and girls is still prevalent. "Children are very literal about colors relating to products," say Selina S. Guber and Jon Berry. Thus, most doll packages are still pink, and purple cough remedy packages still identify "the kids' favorite" grape flavor.

However, for the majority of products targeted at today's children, there is no need to associate specific colors. Most packages of

Courtesy of Dixon Ticonderoga Co.

Exhibit 10.2

Courtesy of Dixon Ticonderoga Co.

Exhibit 10.3

children's products, especially toys, are a riot of colors, on the assumption that the bright colors will stimulate excitement and desire. Even food packages that appeal to the young generation, such as snack chips, should be and frequently are extremely colorful. Primary colors—reds, blues, and greens—dominate on toy packages, while black often serves as the background to frame dramatic figures or emphasize some type of action, with an occasional fluorescent color thrown in to identify "cool" products.

An example of how to use color effectively on packages targeted primarily at children is the redesign of Prang, a line of art supplies marketed by Dixon Ticonderoga Company (see Exhibit 10.2, showing old design). Reacting to the dominance at stationery and art supplies stores of the brightly colored but otherwise boringly designed packages for Crayola markers and crayons, the new Prang packages utilize *black* backgrounds and artistic animal themes. Each product is assigned its own animal, shown in black-and-white, and an example of that animal as if it had been colored by the product contained in the box. This combination of black backgrounds and attractively colored, charmingly rendered animals distinguishes Prang's packages from those of their competitors and forms a dramatic billboard at the point of sale (see Exhibit 10.3, showing new design).

Structures—Structures for toy packages often anticipate the needs for effectively displaying the products or storing them after use. In order to clearly display the products, packages for dolls, licensed characters, car models, and toy trucks are among those products that are most often packaged in *windowed* and structurally complex units. Games are often packaged so that the sometimes numerous play components can be stored in compartments created for them as part of the packaging structure.

However, manufacturers of other products, such as everyday food consumed by kids, like milk, cereals, and chips, have paid little attention to the ergonomic needs of children. Says James U. McNeil, author of *Kids as Customers,* "It's hard not to notice the frequent advertising and publicity directed to children by the fresh milk industry and hard not to notice, also, that the strategy stops at the package. Packaging for kids certainly appears not to be a distinct art among the companies in the kids market." Mr. McNeil goes on to describe the difficulty a young child has opening a large milk carton and pouring from it. Plastic milk containers are no easier for youngsters to manipulate.

A notable exception are the plastic single-serving bottles used by Mayfield Dairy Farms of Athens, Tennessee, which provide a narrower neck for easy holding, a plastic cap for easy opening, and a wider mouth for easier drinking by kids (see Exhibit 10.4). Why, one

Exhibit 10.4

might ask, do not more manufacturers of products that are meant for consumption by kids create purchase interest through kid-oriented packaging structures?

While complicated, packaging for children is probably one of the most exciting and creative marketing areas. Targeted at consumers who progress from childhood through the teens to adulthood, the marketer has limitless opportunities to capture the imagination of current and future customers. The kids of today will be the parents of tomorrow. But to be successful, marketers must be precise as to whom they want to reach and what is most likely to motivate them at what level of intellect. The age, the sex, the household environment, the region in which they live, the urban or suburban location of the target audience—all are critically important elements that will influence the strategic approach of marketing products for the young consumer and of designing packaging that will motivate their purchase interest.

With the assistance of such background research information, and a keen sensitivity to the needs and the behavior of children, it should be possible to offer what few products do: happiness!

Packaging for the Fifty-Plus Market

While the darling of marketers has always been kids—either their own or those who are marketing targets for their products—the attention that marketers give to the other end of our life span, the fifty-plus generation, is quite another story.

As everyone involved in marketing consumer products knows, the target market of most products rarely reaches beyond middle age, in the apparent belief that once consumers have reached their fifties, their active lifestyles end and they become couch potatoes who are no longer interested in acquisitions. We have all had the experience of participating in consumer research planning in which few, if any, consumers beyond their middle forties were considered appropriate research respondents, unless it concerned products specifically targeted to mature audiences, such as pharmaceuticals. This makes little sense in the face of the statistics regarding the lifestyles and the size of the mature population in the United States, as well as in most other economically developed countries.

The facts are that since 1900 the percentage of Americans over sixty-five has more than tripled (4.1 percent in 1900; 12.7 percent in 1993). The most rapid increase is expected from 2010 to 2030. The American Census Bureau estimates that the number of Americans over age fifty will soar 80 percent, past 100 million, to account for one-third of the entire U.S. population by the year 2025. In marketing terms, that's right around the corner. This trend is also occurring in western Europe and Japan, where one in five citizens in industrial countries will soon be over age sixty-five.

Advances in medical care and changes in lifestyle are resulting in the healthiest and wealthiest mature population in history. In the United States, the fifty-plus group has as much discretionary income as all other age groups *combined*—adding up to approximately $150 billion. That's a big slice out of the marketing pie!

While marketers and ad agencies continue chasing the younger market, they have not taken a recent look at the circulation of magazines. Had they done so, they would have noticed that in the United States *Modern Maturity* magazine is by far the surprising leader. With a circulation of more than twenty-two million, the official publication of the American Association of Retired Persons (AARP) is way ahead of the pack. Even publishing giants like *Reader's Digest* (circulation sixteen million) and *TV Guide* (just under sixteen million readers) can't hold a candle to *Modern Maturity,* targeted to readers age fifty and over!

These statistics should give a clear indication to marketers of retail products that the fifty-plus consumer group deserves every bit as much respect as under-fifty consumers and that it would make sense to pay more attention to seniors' needs as purchasers and users of packaged products. From a marketing strategy point of view, everything you read and hear bears witness to the fact that senior consumers see themselves as young and that this consumer segment constitutes an important and growing market. In spite of all this evidence, packaging continues to be focused almost exclusively on the needs and desires of younger consumers. When structural and graphic package design issues are on the table, little attention is being paid to the declining manual and ocular faculties of the fifty-plus consumer.

However, it is a mistake to lump the fifty-plus consumers together into one group. The United States Census Bureau segments the mature market into several age groups, of which two should be of particular interest to marketers of retail products:

- ages 50 to 64, when consumers are at their earning peak while household- and family-related expenses decrease as children leave home and houses are paid off

- ages 65 and over, when most consumers are retired and pursuing new interests for which there previously was no time

When strategizing package design, marketers should be aware of some of the ergonomic issues that affect consumers who fit into these age brackets. Almost 50 percent of the population over the age of sixty-five has some form of arthritis, and more people of all ages are encountering difficulties using products that have not been designed with differences in ablehandedness in mind. Dr. Margaret Wylde of the Institute of Technology Development notes, "Grip strength among women is half that of men. We have the greatest strength in our hands during our twenties." Dr. Wylde goes on to say, "Grip strength in normal hands decreases steadily as we age. . . . It is only half that of our youth when we reach our seventies and eighties. Far too many products create excessive demands on hands."

Research conducted over a period of several years by Interbrand Gerstman+Meyers, an international brand identity and package design consultancy, delved into this issue in great detail. Based on a series of focus group interviews with consumers aged fifty to seventy-five, complaints about the structural and visual aspects of packaging cropped up repeatedly. Things like child-resistant caps for pharmaceutical products that resist opening efforts by the elderly, "push here" paperboard cartons that refuse to yield, wide-mouthed jar lids that are difficult to grasp, and copy too small to read without the help of a magnifying glass were all frequent targets for complaints by the mature respondents.

Delving into these concerns in great detail, the research pinpointed a number of interesting issues that impact directly on the structural and graphic design of packages. These issues are sufficiently critical that they should play a significant role in package design planning for any consumer product, particularly when enhanced packaging features that offer advantages to mature consumers can attract younger consumers as well.

Visual Issues for the Fifty-Plus Consumer

Visual packaging issues for the fifty-plus consumer are numerous and enlightening.

- By age sixty-five, ability to focus, resolve images, distinguish among colors, and adapt to different lighting conditions diminishes. Increased opacity of the eye, including cataracts, and clouding of the lens of the eye reduce the amount of light that enters. This makes the contrast level of any visual material a critical design element.

- The diminished ability to focus impairs the perception of small forms and letters. Therefore, small, crowded type, as found on many small pharmaceutical labels and packages, as well as other small packages, can be difficult to read.

- The small size of secondary type, such as usage instructions and ingredients, frustrates senior citizens in their quest to obtain maximum information about the products they purchase (see Exhibit 10.5). Respondents report using a support system of tools such as a magnifying glass and strong light in addition to their bifocals to help read packaging copy. They want type to be large enough to read without a magnifier and arranged in more readable formats. They find small blocks of copy and short lines easier to read than copy arranged on long, extended lines.

- The stronger the contrast between the colors of text and background, the easier it is to find and read important text. Interestingly, men and women sometimes differ in this respect. While men prefer optimal contrast—such as black type on white backgrounds—women lean more toward softer background colors, even at the risk of slightly reducing contrast.

Type styling for pharmaceutical packages

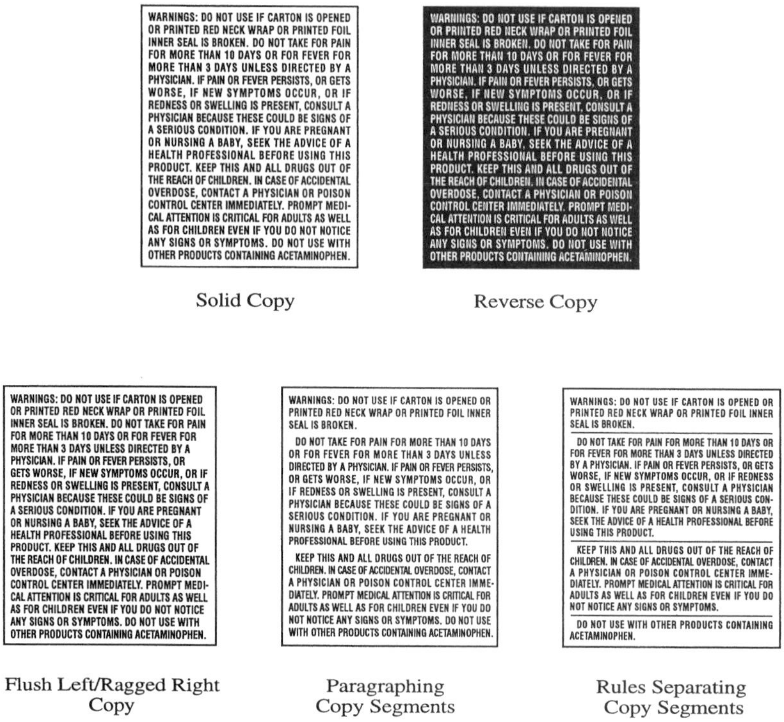

WARNINGS: DO NOT USE IF CARTON IS OPENED OR PRINTED RED NECK WRAP OR PRINTED FOIL INNER SEAL IS BROKEN. DO NOT TAKE FOR PAIN FOR MORE THAN 10 DAYS OR FOR FEVER FOR MORE THAN 3 DAYS UNLESS DIRECTED BY A PHYSICIAN. IF PAIN OR FEVER PERSISTS, OR GETS WORSE, IF NEW SYMPTOMS OCCUR, OR IF REDNESS OR SWELLING IS PRESENT, CONSULT A PHYSICIAN BECAUSE THESE COULD BE SIGNS OF A SERIOUS CONDITION. IF YOU ARE PREGNANT OR NURSING A BABY, SEEK THE ADVICE OF A HEALTH PROFESSIONAL BEFORE USING THIS PRODUCT. KEEP THIS AND ALL DRUGS OUT OF THE REACH OF CHILDREN. IN CASE OF ACCIDENTAL OVERDOSE, CONTACT A PHYSICIAN OR POISON CONTROL CENTER IMMEDIATELY. PROMPT MEDICAL ATTENTION IS CRITICAL FOR ADULTS AS WELL AS FOR CHILDREN EVEN IF YOU DO NOT NOTICE ANY SIGNS OR SYMPTOMS. DO NOT USE WITH OTHER PRODUCTS CONTAINING ACETAMINOPHEN.

Solid Copy

WARNINGS: DO NOT USE IF CARTON IS OPENED OR PRINTED RED NECK WRAP OR PRINTED FOIL INNER SEAL IS BROKEN. DO NOT TAKE FOR PAIN FOR MORE THAN 10 DAYS OR FOR FEVER FOR MORE THAN 3 DAYS UNLESS DIRECTED BY A PHYSICIAN. IF PAIN OR FEVER PERSISTS, OR GETS WORSE, IF NEW SYMPTOMS OCCUR, OR IF REDNESS OR SWELLING IS PRESENT, CONSULT A PHYSICIAN BECAUSE THESE COULD BE SIGNS OF A SERIOUS CONDITION. IF YOU ARE PREGNANT OR NURSING A BABY, SEEK THE ADVICE OF A HEALTH PROFESSIONAL BEFORE USING THIS PRODUCT. KEEP THIS AND ALL DRUGS OUT OF THE REACH OF CHILDREN. IN CASE OF ACCIDENTAL OVERDOSE, CONTACT A PHYSICIAN OR POISON CONTROL CENTER IMMEDIATELY. PROMPT MEDICAL ATTENTION IS CRITICAL FOR ADULTS AS WELL AS FOR CHILDREN EVEN IF YOU DO NOT NOTICE ANY SIGNS OR SYMPTOMS. DO NOT USE WITH OTHER PRODUCTS CONTAINING ACETAMINOPHEN.

Reverse Copy

WARNINGS: DO NOT USE IF CARTON IS OPENED OR PRINTED RED NECK WRAP OR PRINTED FOIL. INNER SEAL IS BROKEN. DO NOT TAKE FOR PAIN FOR MORE THAN 10 DAYS OR FOR FEVER FOR MORE THAN 3 DAYS UNLESS DIRECTED BY A PHYSICIAN. IF PAIN OR FEVER PERSISTS, OR GETS WORSE, IF NEW SYMPTOMS OCCUR, OR IF REDNESS OR SWELLING IS PRESENT, CONSULT A PHYSICIAN BECAUSE THESE COULD BE SIGNS OF A SERIOUS CONDITION. IF YOU ARE PREGNANT OR NURSING A BABY, SEEK THE ADVICE OF A HEALTH PROFESSIONAL BEFORE USING THIS PRODUCT. KEEP THIS AND ALL DRUGS OUT OF THE REACH OF CHILDREN. IN CASE OF ACCIDENTAL OVERDOSE, CONTACT A PHYSICIAN OR POISON CONTROL CENTER IMMEDIATELY. PROMPT MEDICAL ATTENTION IS CRITICAL FOR ADULTS AS WELL AS FOR CHILDREN EVEN IF YOU DO NOT NOTICE ANY SIGNS OR SYMPTOMS. DO NOT USE WITH OTHER PRODUCTS CONTAINING ACETAMINOPHEN.

Flush Left/Ragged Right
Copy

WARNINGS: DO NOT USE IF CARTON IS OPENED OR PRINTED RED NECK WRAP OR PRINTED FOIL INNER SEAL IS BROKEN.

DO NOT TAKE FOR PAIN FOR MORE THAN 10 DAYS OR FOR FEVER FOR MORE THAN 3 DAYS UNLESS DIRECTED BY A PHYSICIAN. IF PAIN OR FEVER PERSISTS, OR GETS WORSE, IF NEW SYMPTOMS OCCUR, OR IF REDNESS OR SWELLING IS PRESENT, CONSULT A PHYSICIAN BECAUSE THESE COULD BE SIGNS OF A SERIOUS CONDITION. IF YOU ARE PREGNANT OR NURSING A BABY, SEEK THE ADVICE OF A HEALTH PROFESSIONAL BEFORE USING THIS PRODUCT.

KEEP THIS AND ALL DRUGS OUT OF THE REACH OF CHILDREN. IN CASE OF ACCIDENTAL OVERDOSE, CONTACT A PHYSICIAN OR POISON CONTROL CENTER IMMEDIATELY. PROMPT MEDICAL ATTENTION IS CRITICAL FOR ADULTS AS WELL AS FOR CHILDREN EVEN IF YOU DO NOT NOTICE ANY SIGNS OR SYMPTOMS. DO NOT USE WITH OTHER PRODUCTS CONTAINING ACETAMINOPHEN.

Paragraphing
Copy Segments

WARNINGS: DO NOT USE IF CARTON IS OPENED OR PRINTED RED NECK WRAP OR PRINTED FOIL INNER SEAL IS BROKEN.

DO NOT TAKE FOR PAIN FOR MORE THAN 10 DAYS OR FOR FEVER FOR MORE THAN 3 DAYS UNLESS DIRECTED BY A PHYSICIAN. IF PAIN OR FEVER PERSISTS, OR GETS WORSE, IF NEW SYMPTOMS OCCUR, OR IF REDNESS OR SWELLING IS PRESENT, CONSULT A PHYSICIAN BECAUSE THESE COULD BE SIGNS OF A SERIOUS CONDITION. IF YOU ARE PREGNANT OR NURSING A BABY, SEEK THE ADVICE OF A HEALTH PROFESSIONAL BEFORE USING THIS PRODUCT.

KEEP THIS AND ALL DRUGS OUT OF THE REACH OF CHILDREN. IN CASE OF ACCIDENTAL OVERDOSE, CONTACT A PHYSICIAN OR POISON CONTROL CENTER IMMEDIATELY. PROMPT MEDICAL ATTENTION IS CRITICAL FOR ADULTS AS WELL AS FOR CHILDREN EVEN IF YOU DO NOT NOTICE ANY SIGNS OR SYMPTOMS.

DO NOT USE WITH OTHER PRODUCTS CONTAINING ACETAMINOPHEN.

Rules Separating
Copy Segments

Exhibit 10.5

- On the positive side, mature consumers are comfortable with the updated nutrition labels and understand and utilize the information they provide. Seniors often take the time to read the ingredients of the food products they purchase. They are concerned about health and appreciate the new, reformatted nutrition labels that aid them in purchasing food with ingredients appropriate to their needs.

- Due to mature consumers' emphasis on health, they are very concerned about expiration dates and suspect that hard-to-read and distorted date imprints indicate the marketer's attempt to conceal the expiration information. Similar in context, mature consumers expect packaging graphics to be a true representation of package contents and are upset if they find information or visuals on the package to be misleading. They are also angered and feel they are being cheated when packages appear to be the same size but contain different amounts of the same product.

- Mature consumers look for cues such as color, print style, and graphics that they are familiar with and have been exposed to through advertising and packaging. They use these cues to help locate the product they want. In contrast, changes in packaging structures or graphics that are too drastic can diminish familiar cues and make it harder for customers to find the product they want.

- Cues that effectively delineate, highlight, or emphasize important product information on package back panels can facilitate readability. Enlarged size of the text, paragraphing, rules separating various copy elements and a change of color for such wording as *warning* and *do not* help consumers of any age to find and comprehend important information.

- The use of icons can facilitate understanding of product use and preparation, although icons alone do not usually communicate sufficient information for many senior consumers. A combination of icons and brief written instructions is an effective method of accenting significant product features and supporting important details regarding product use.

- Senior consumers react positively to package photographs, especially on food packaging. Photographs identify the product and the flavor, suggest ingredients, create an image, and help consumers determine if there is further interest in the product. Once this interest has been generated, then nomenclature, such as flavor descriptors, fat, sodium and cholesterol contents, net weight, and expiration dates are examined in detail.

- Despite the older consumers' problem with distinguishing certain colors, color coding does often help them to identify certain product categories (such as green for decaffeinated coffee) and verify product varieties for which they are looking. But color is often not their first cue of identification. Photographs on the packages are more meaningful to them than the color coding. Color coding becomes more of an issue with repeat purchases. Respondents remember package colors and will look for these to ensure that they are buying the product they want.

Structural Issues for the Fifty-Plus Consumer

Ergonomic issues—the facility of holding, opening, dispensing product from, and reclosing packages—are even more critical in the marketing of products to the fifty-plus consumer segment.

- All things being equal (product, size, value, and so on), many mature consumers are willing to try different products if a more appealing package graphic or an improved structure catches their attention and interest. However, packaging structures must offer a significant improvement over current alternatives to increase the value of the products inside to the fifty-plus consumer. Structural innovations that *really solve* ergonomic problems are appreciated, while structures or features that do not are considered to be promotional frills that are not necessarily worth their cost.

- Mature consumers are extremely value conscious. They want containers to easily dispense *all* their contents to avoid waste. The desire to fully empty a package and the ability to use up *all* of the product are often identified as important to senior consumers. The tapered neck on some jelly or condiment jars and, of course, on the proverbial ketchup bottle that resists discharging all its contents, even with a knife, are always on the seniors' packaging hate list. Also, they expect packages to last the life of the product without necessitating transfer to an additional container.

- Portion packs are considered to be a benefit by seniors for products not used frequently or in large quantities. They are perceived as preserving freshness for extended periods of time. While there is an expectation to pay more for portion-packed products, the cost is weighed against the perceived benefit.

- Excessive packaging is considered to be misleading and wasteful unless it provides product protection or information that aids in product selection.

- Because mature consumers are concerned about dropping things and the possibility of resulting breakage, they prefer plastic containers over glass. Also, for many products, clear packaging is preferred over opaque so that the contents can be viewed prior to purchase and checked for broken or crushed product. The ability to view the product is especially important to mature consumers in the fresh meat, vegetable, and baked goods categories.

- While opening packages can sometimes be a challenge to senior consumers, they acknowledge the necessity for closures that protect products from spoilage and for tamper-proof seals to ensure healthy contents. However, there is a need to strike a balance. Seniors seem to be willing to put effort into initial opening, but subsequent openings and closings must be more easily achieved or these devices will become counterproductive. For example, analgesics packages that require the exact lining up of arrows before the caps can be removed cause seniors to pry them open with instruments, disabling their safety mechanisms. The "Fast Cap" on analgesics marketed by Tylenol and Arthritis Foundation brands, for example, provides a welcome alternative to opening problems, even though the products contained in such bottles were marketed for many years before the manufacturers recognized the need for easier product access (see Exhibit 10.6).

- Handles that allow senior consumers to achieve a strong grip with leverage for pouring are most desirable for

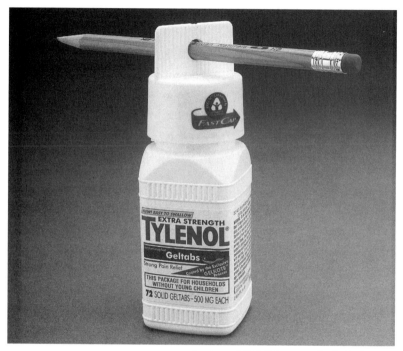

Courtesy of McNeil Consumer Products Company

Exhibit 10.6

large containers containing liquids. For example, the
"Easy Grip" handle feature on the Motts apple juice
plastic bottle (see Exhibit 10.7) provides better leverage
than the hard-to-balance glass containers that have to be
held and tilted by their necks. Similarly, the large handles
on many plastic detergent bottles are welcome features
for seniors, as well as all other consumers.

- There are numerous other types of convenience features
  to which the mature consumer, as well as consumers of
  all ages, are likely to respond favorably, such as ring tabs
  that facilitate removal of a metal lid (though seniors
  would like them to accommodate *two* fingers, instead of
  one, to facilitate easier pull); perforations on the backs of
  blister packs for easy opening; milk and juice containers
  with screw caps; tear tapes on plastic film overwraps; and
  cereal, flour, pasta, chips, and cookie bags with zipper
  features that make them easier to open and reseal.

It should be obvious by now that paying closer attention to the manual and ocular requirements of men and women aged 50 and older will help marketers of retail products develop packaging that appeals to this substantial and growing consumer segment. Considering the significant trend of the expanding population of mature consumers and marketers' increasing urgency to respond to the need for achieving a competitive advantage, it seems that the time is appropriate to step back and reassess how this market segment can be better served. In that endeavor, packaging can play a central role.

Even more significant in terms of potential business is the indication that packaging features that meet the needs of the mature market segment will almost certainly be equally welcomed by the general population. Marketers who make this philosophy the cornerstone of their design development strategies for *all* consumer packages will very likely gain not only the endorsement of the segment of our population identified as the fifty-plus market but the enthusiasm of consumers of *all* ages.

Courtesy of Mott's USA

Exhibit 10.7

11 The Technology of Packaging

Although the production of packaging is rarely the direct responsibility of marketing executives, the overall concern for the effectiveness of packages at the point of sale cannot be ignored by them. The fact is that the effect of every step on the road of package design and development, whether it is appearance, functionality, or production, will impact the ultimate success or failure of the product. For that reason, basic familiarity with some of the key technical aspects of package development, such as standard packaging forms and materials, the most frequently used methods of package printing, and what happens when the approved design concept is prepared for production, should be in the vocabulary of everyone involved in the package design development process.

An important step that is often undervalued or acted upon belatedly by marketing executives is to arrange for *pre-production* coordination between designer, internal packaging engineers, and external packaging suppliers.

Pre-Production Coordination

Pre-production coordination involves meetings by all those responsible for packaging specifications and development, purchasing, and packaging implementation. For package development to proceed smoothly and to succeed in the marketplace, it is essential to arrange for such a meeting as early as possible in the design development program. The packaging requirements of the manufacturer and the production capabilities, equipment, and production procedures of each supplier may vary substantially. These differences impact directly on the design flexibility and, therefore, must be taken into account. Discussing these issues prior to production can avoid complications and costly delays during the final package production process.

Designers are expected to develop unique, new concepts that support the marketing position of the product. These may occasionally include ideas that do not adhere to existing production procedures or are not geared to the production capabilities of suppliers selected by the manufacturer's purchasing department. This possibility emphasizes the importance of reviewing production parameters during the early stages of design development, or even *before* the start of the design development, so that the design consultant will be informed about equipment, product particularization, and cost parameters. This information may identify important issues such as the preservation and protection of the products or the interaction of packaging materials with the products, especially if it involves foods, liquids, pharmaceuticals, chemicals, fragile products, and products sensitive to light or atmospheric conditions.

If a new or modified packaging structure has been created, the manufacturer who will produce the packages—whether cartons, bags, trays, bottles, or any other type of container—should be given the opportunity to interact with the designer *early* in the project. No matter how much experience the designer may have with materials, production, and printing of packages, early discussion will frequently result in suggestions by the supplier that will be beneficial to the designer, including opportunities that the designer may have overlooked.

Packages for products marketed in large quantities may be produced by several printers and at different locations so that production artwork has to satisfy the preprint and printing procedures of these different suppliers. The same holds true when product lines consist of a variety of packaging forms. A large line of household gadgets may be packaged in blister packs, plastic bags, folding cartons, clam packs and several other structures, each produced by different package suppliers. A line of beverages may be available in aluminum cans with dry offset printing, plastic bottles with flexographically printed paper labels, glass bottles with screen printing, and paperboard multipacks printed by the lithographic process. All of these require different preparation methods—yet all of them must culminate in a coherent family of packages and communicate in a visually cohesive manner.

Thus, pre-production meetings that provide an opportunity to discuss all production requirements invariably result in time and cost savings by minimizing misunderstandings or misinterpretation. Although it is sometimes difficult to get all the parties together at one location, prepress meetings should never be neglected or bypassed for reasons of expediency. The time and effort spent for prepress coordination will pay for itself by preventing the need for correcting costly mistakes or jeopardizing the schedule of the product launch.

Scheduling Package Design and Production

Another important facet of the package design development process is having a clear understanding of the time increments required for the development of packages. Once the decision for launching a new product or revitalizing an existing one has been made, marketing management, understandably, is always anxious to move package development forward expeditiously.

But it is always prudent to look before you leap. Walter Soroka in his book *Fundamentals of Packaging Technology*, explains "Packaging is an extraordinarily complex endeavor that must be viewed as a part of a larger system, within which every activity has some impact or demand on the package." The demands, he continues to point out, are often not mutually compatible, so that the production system—and that includes the time required to accomplish each of the production components—directly affects the final outcome of the package.

Time schedules naturally vary substantially from project to project. The time span of a package development program from its initiation to the final production can vary dramatically, ranging from a few weeks or months to several years. Without even counting the time required for solving marketing issues, the technical complexity of the packages, such as the accessibility of materials, the time needed for package testing procedures, the number of packaging components, and the condition of existing plant equipment or the time for installing new equipment, require establishing a realistic timetable for the entire package development project as early as possible. Last but not least in considering time schedules is the experience of the technical personnel, as well as the efficiency of the suppliers.

Because of the complexity of these multi-layered requirements, it is impossible to generalize timetables or suggest an infallible procedure to expedite the technical steps during the course of package development. Even if this were feasible, the ever present potential for changes in marketing or technical issues while the program is in process adds another dimension that makes any attempt to generalize package development schedules a risky assumption.

However, similar to developing annual sales plans that act as a *target* to shoot for, it is essential for marketers to *target* a realistic tracking schedule for package development. The hypothetical example of a package development schedule for a beverage package (shown in Exhibit 11.1) is representative of the type of a flowchart that shows the time allocations for each step in the

Package Design Progress Schedule

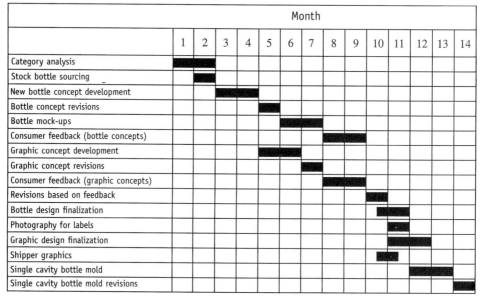

	Month													
	1	2	3	4	5	6	7	8	9	10	11	12	13	14
Category analysis	■	■												
Stock bottle sourcing		■												
New bottle concept development			■	■										
Bottle concept revisions					■									
Bottle mock-ups						■								
Consumer feedback (bottle concepts)								■						
Graphic concept development					■	■								
Graphic concept revisions							■							
Consumer feedback (graphic concepts)								■						
Revisions based on feedback										■				
Bottle design finalization											■			
Photography for labels											■			
Graphic design finalization											■	■		
Shipper graphics										■				
Single cavity bottle mold												■	■	
Single cavity bottle mold revisions														■

Courtesy of Walter Soroka, *Fundamentals of Packaging Technology,* Institute of Packaging Professionals

Exhibit 11.1

development procedure. While the requirements will vary from project to project, the establishment of such a flowchart with input from designer, engineers, and suppliers is recommended for every package development project.

What Marketers Should Know About Package Construction and Materials but Are Afraid to Ask

While most marketing executives are familiar with the general types of packages required for their products—whether folding cartons, glass bottles, plastic containers, or blister packs—they are not likely to have in-depth familiarity with the technical specifications of each of these. The truth is that this is not necessarily a marketing responsibility, as most marketers have access to the assistance of internal engineering personnel or outside consultants. Nevertheless, a general understanding of the basic materials available to fill, store, and ship their products as well as of some of the structural and printing procedures that will deliver the desired visual results will be useful in expediting package design and making development decisions.

To begin with, there are different needs for different types of packages. These may include

- primary packages—the packages that contain the actual product

- secondary packages—outer packages containing one or more primary packages for reasons of protection or visual appearance, such as a cereal carton with an inner bag to keep the cereal fresh, a multipack for several beverage cans or bottles, or a perfume package to protect and further enhance the bottle inside

- distribution packages—packages, such as corrugated cartons or palletizing shrink-wraps, used for the protection and shipping of the primary and secondary containers

Unfortunately, packaging requirements often do not offer an unlimited range of choice of materials or packaging forms. Cost considerations and the existence of in-plant equipment often limit the choice of packaging forms and materials to those readily available. But even when this is the case, an analysis of available resources must be made in connection with every package design development, whether it concerns a new product, the change of a package feature, the addition of a product variety or flavor to an existing brand line, or a simple graphic change.

This analysis is especially important when glass or plastic containers are involved, requiring a critical analysis of materials, structures, volume, closures, and packaging line filling procedures. There are also concerns about palletizing efficiencies, top load strength during shipping and storage, and potential damage during shipping. Similarly, paperboard containers—whether folding cartons, setup boxes, carded packages, or corrugated containers—require precise specifications and operational procedures regarding materials, protective barriers, carton erection, folding, gluing, and closures, as well as stacking, shipping, and palletizing. Last but not least, film bags and pouches, as well as metal containers, such as three-piece steel cans, drawn aluminum cans, and aerosol containers, offer a whole range of shapes, materials, and lamination alternatives that are subject to the same scrutiny as other container forms.

At the retailer level, packages have yet another set of physical requirements: they must fit the limitations of shelf height and depth and other space requirements at various stores. If these requirements vary substantially between, for example, a hardware store and a mass merchandiser, a compromise will have to be worked out to fit these retail conditions or *different packaging forms* for several retail outlets may have to be considered. For example, when Wal-Mart stores demanded that Eveready batteries, ordinarily available in blister packs, be shrink-wrapped to save space in their stores, Eveready Battery Company had no choice but to oblige or risk losing this key customer.

When the products are food, the shelf life of the products during storage, shipping, and at retail play a major role in the choice of materials and packaging forms. Again, there is a strong possibility that some types of food will utilize a variety of packaging forms. Some beverage brands, for example, are available in glass, plastic, and aseptic packages, each with different engineering requirements regarding materials, construction, filling and packaging line procedures, stacking strength, shelf life, and shelf placement limitations. Other concerns for food marketers are the environmental conditions in some areas. The climatic conditions in Miami and Minneapolis during the same seasons are quite different. This is not to mention the sharply divergent climatic considerations facing products that are marketed internationally,

When all is said and done, all of these diverse factors are subject to one overriding consideration: the bottom line. The cost of packaging is relative to the cost of the products, the types of products, and even the cost of the products and packaging of other brands in the category. A change in package design, no matter how slight, or the redesign of packages for an entire product line can have substantial effect on the product's annual sales volume and profits. The risk of a package change or development of a new package must always be weighed by the marketer against the benefit expected from such a move. Each of the many technological facets of package design—type of container, choice of material, material cost, product protection, palletization, manufacturing process, shipping requirements, distribution channels, scheduling limitations, and manufacturing costs—all must be analyzed and considered carefully by marketing and technical personnel before final package design specifications can be agreed upon.

The Most Popular Printing Methods for Packaging

It may not be necessary for marketing executives to be thoroughly knowledgeable about the intricacies of the many printing methods available to reproduce packaging graphics. Nevertheless, just as in connection with the physical properties of packaging structures, a basic overview of the printing methods currently used on the majority of packages should be helpful.

Printing methods used for package reproduction are generally divided into three basic categories:

- relief printing—flexography, letterpress, offset flexography (dry offset)
- planographic printing—offset lithography
- recess printing—gravure, rotogravure

Brief descriptions of some of the primary printing methods for packaging may help marketers to better evaluate the opportunities and limitations for each:

Flexography is a printing method using relief rubber or polymer plates. Raised images transfer ink (sometimes water based) to the substrates. Flexography, a relatively cheap printing method, is used most often on roll-fed paper, paperboard, corrugated board, and film. Advantages include low cost and the ability to apply printed images to flexible and relatively uneven surfaces. Its disadvantage, vis-à-vis lithography, is primarily the need for coarser print screens. This is especially true when printing photographic subjects. However, strides have recently been made to achieve greatly improved results, making flexography a popular, low-cost printing process.

Offset lithography is the preferred printing method used for decorating paper labels and cartons. This printing method uses the immiscibility of oil-based ink and water to transfer an image from a metal plate onto a rubber blanket that in turn *offsets* the image to the substrate. Each color requires a separate set of printing elements and printing stations (hence limiting available colors). Advantages of lithographic printing include the richness and range of available color combinations. This is especially noticeable when printing photographic subjects. Lithographic reproduction of photography is created through the use of halftone screen plates and process color inks (yellow, magenta, cyan, and black). Dark colors are sometimes used in place of black. Custom-matched colors, if required in addition to standard process colors, require additional printing plates and printing stations and the cost associated with these.

Offset flexography (dry offset) is a method used for printing on metal cans, plastic tubs, and cups. This method uses a series of small, rotating cylinders in a circular, carousel-like arrangement that transfer images simultaneously from an offset blanket to rotating cans or tubs (see Exhibit 11.2). Advantages include speed and low printing cost. Disadvantages result from simultaneous printing of wet inks, requiring total separation of colors to prevent color degradation. This restricts images to those using flat colors, though some designers are finding clever ways of creating images that get around these limitations.

Letterpress is similar to flexography. Letterpress uses relief plates for transferring images to substrates but utilizes oil-based inks. Letterpress, once utilized extensively for packaging, is no longer one of the preferred methods for packaging reproduction but is still used for printing roll-fed pressure-sensitive labels.

Gravure is a method using cylinders (hence *roto*gravure) into which images are engraved or etched resulting in wells containing ink. Using roll stock, the ink is transferred to paper and paperboard through suction resulting from the pressure of the cylinder on the substrate. Rotogravure printing is used extensively for very long print runs, since it can process labels and cartons at very high speeds and uses metal cylinders that resist wear during the long runs. Advantages include speed and cylinder longevity resulting in lower production costs, but startup costs are usually higher because of the investment in more expensive cylinder preparation.

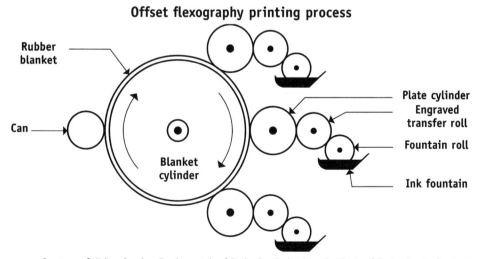

Offset flexography printing process

Courtesy of Walter Soroka, *Fundamentals of Packaging Technology,* Institute of Packaging Professionals

Exhibit 11.2

Heat transfer is a method of using heat to transfer a thermoplastic ink image from a carrier film to a substrate, used mostly for plastic containers. Advantages include the ability to apply color images to soft plastic substrates, achieving a variety of visual effects. This method is particularly popular for containers in the health and beauty aid category.

Hot stamping is a method that like heat transfer printing, utilizes a carrier film, thermoplastic ink, and heat to transfer images onto plastic containers or film. The difference lies in that the image is transferred by a heated die pressed against the carrier film that transfers the image to the substrate. Advantages are the ability to achieve shiny surfaces and images that mimic aluminum, silver, and gold.

Screen printing is a method of transferring color images to glass or plastic surfaces by forcing ink through a metal or plastic screen onto the substrate. The ability to obtain a thick and lasting layer of color on the substrate is particularly advantageous for such containers as returnable glass beverage bottles and perfume bottles. A disadvantage is in its being restricted to a limited number of colors, since the screening process is relatively slow, reflecting on the package production cost.

As always in package design, generalizations are risky. In that sense, this list of printing methods is not complete nor should it be regarded as definitive. There are many possible variations and combinations of printing methods that could be explored with the design consultant and the technical experts on your staff. But used as a general guide, the list of alternate print production methods should be helpful.

Pre-Production Preparation

Pre-production preparation for printing, once called *mechanical artwork*, is done almost entirely electronically today. While a few smaller manufacturers and printers still require traditionally prepared paste-ups of reflective artwork—that is, text and other visual material pasted on a flat board in actual printed size—the vast majority of printers today are set up for processing artwork prepared by the designer electronically using computer disks.

In order to prepare production artwork after the design has been approved, the marketer, the design consultant, and the supplier or suppliers confer to determine the extent of the pre-production preparation required. These discussions review the number of SKUs for which artwork has to be prepared, the method of art preparation preferred by the printers, printing methods to be used, printing station availability, pre-production costs, and any other technical details pertaining to the production of the packages.

There are a number of elements that must be supplied by the marketer and the printer before the designer begins the preparation of production artwork:

- The marketer must supply final text, including product and promotional copy; nutritional, ingredient, and caution copy; the U.P.C.; and any other copy or identification elements.

- The printer must supply die specifications, both in electronically prepared format (which the designer will use to prepare final production artwork) and traditional die imprinted boards (to double-check the correctness of final production art).

One of the characteristics of computer-aided design development is that much of the artwork needed for print production has already been anticipated during the concept development. When the designer prepares mock-ups for the marketer's final design approval, the actual text and the final photographs or illustrations are usually included on the mock-ups. It is important for the marketer to realize, however, that this artwork, as realistic and finished as it may appear, is rarely if ever directly usable for print production. Though the designer is aware of the printability of the concepts, text on these mock-ups may not be final nor proofread, and the colors used during concept development were selected by the designer from the extensive color palette available on the computer and intended to simulate the final package appearance for presentation only. For that reason, the purpose for pre-production preparation is to *convert* the design files into reproducible artwork, including the exact specifications of matched colors and process tint colors for halftone reproduction.

After the production art files have been completed, a reading stat is prepared from the production art and sent to the marketer for approval. The reading stat contains final text, which the marketer and often the marketer's legal department carefully examine, making changes and corrections and returning it to the design consultant for further processing. The final step in the preparation of production artwork by the designer occurs when a full-color electronically produced pre-production printout is sent to the client who is responsible for final approval of the artwork. After the client's approval, the printer will adjust the artwork to specific tolerances for color trapping and bleeds.

Occasionally, though happily very rarely, a last-minute change is desired by the marketer or a copy error or misspelling is discovered *after* the final printout has already passed inspection and has been approved. It must be understood by the marketer that once the final printout has been approved for printing, the costs involving additional services by the design consultant, the printer, or both will be the responsibility of the marketer, no matter where and how an error originated. The possibility of a misunderstanding or ill feelings resulting from such an occasion puts particular pressure on the responsible marketing executive to establish a precise approval process.

Preparation for structural finalization and the approval path connected with it are substantially different from that used in preparing printed material. To begin with, the specifications for a packaging structure are so diverse that they consist of *groups* of documents rather than a single document like the pre-production printout used for approval of graphics. Technical documents in connection with structural package specifications relate to every critical package performance factor: materials, dimensions, component specifications, glue or adhesive specifications, stress tolerances, volume, weight, machinability, and production costs, depending on the types of packages.

Because of these complexities and the need for interfacing with many other departments, particularly plant production personnel, responsibility for handling these steps usually passes from the design consultant to the manufacturer's package development department soon after the design concept has been approved for production. The consultant's direct involvement often ends with the preparation of an electronic file for the production of prototype packages.

If the package structure is for a bottle or another type of glass or plastic container, one or more single-cavity molds are prepared from the electronic file. These molds are usually prepared by the designated supplier in coordination with the marketer's production staff for the purpose of testing the performance of the package with regard to handling, stress-related factors, machinability, packaging line compatibility, and so forth. Assuming that the single-cavity molds perform to expectations, or after various complications have been ironed out and the single-cavity mold has been fitted to the manufacturer's packaging system, it will then be approved for full production.

Through all these steps, even when handled by the marketer's internal engineering staff, the project will benefit from your keeping the design consultants apprised and involved enough to benefit from the consultants' experience and interest in seeing their efforts bear the fruits of a successful packaging unit.

Production and Printing Follow-Up

"It ain't over until it's over!" This memorable quip by Yogi Berra is as applicable to package design development as it is to baseball games.

Never take for granted that the packages, no matter how carefully and expertly prepared, will automatically come out exactly right. Supervision during the manufacturing process, printing process, or both is as critical as every other step from design to finalization procedures.

In a large corporation, it is likely that there are specialists responsible for packaging manufacture and printing. These specialists supervise production runs and give final approval of the packages.

Some companies require specialists of the design consultant's staff to participate, together with the marketer's technical experts, in the approval process during a first production run, and designers are usually eager to do so. Participating in the first production run gives the designer a final opportunity to affect the package design development. Some design firms have technical specialists on staff who have in-depth knowledge of packaging manufacture and printing technologies. They speak the same language as the suppliers' technical personnel and, working hand in hand with them, can suggest adjustments in structural or printing production details to achieve the best results.

This very last step in the package design process—the manufacture or printing of the packages—should be considered an integral and critical ingredient for consummating the strategic objective of optimizing the effectiveness of the packages. From the very first meeting to the final production of the packages, the design consultant's integration in the package development process should be a valuable ingredient for achieving optimal results.

12 Packaging and the Law

The United States has been described as one of the most litigious societies in the world. Experienced packaging professionals understand that they have to contend with the pressures and demands of legal requirements as part of their business procedures and that keeping abreast of changes is essential.

Legal issues in package development relate primarily to three major areas:

- laws and regulations regarding container structures (including materials, strength, toxicity, product protection, tampering, environmental issues)
- laws and regulations pertaining to graphics (including trademark, trade dress, copy, claims, contents description, mandatory copy)
- legal agreements between the marketer and the design consultant

The purpose of this chapter is to provide marketers and designers with an overview of some of the legal issues of which they should be aware when introducing a new package or updating or repositioning an existing one. This discussion is not intended to replace professional legal advice. Designers and marketers need to work closely with the legal counsel available to them in order to preclude potentially costly errors in judgment and disruptions to their projects.

Keeping Up with the Law

Packaging law encompasses a wide range of subjects, all of which relate to creating packages containing salable products. Marketers and designers need to familiarize themselves with these matters, not as lawyers but as professionals whose work is linked with the law. Unfortunately, there exists no single source of information regarding packaging laws. Marketers and their design consultants are forced to scan numerous legal references to obtain even a basic overview of the multiple legal issues relating to each package they mutually develop.

Some of the more critical subjects, especially those for packaging of food, beverages, health care products, and chemicals, encompass a wide range of legal compliance issues:

- *structural requirements,* such as tamper evidence, child-resistance, and hazardous material regulations

- *technological issues,* such as material strength, odor transfer, oxygen barriers, chemical migration, corrosion, deterioration, and many more

- *environmental considerations* relating to solid waste disposal or other disposition of packaging materials, particularly glass, plastics, and metals

- *visual regulations* relating to size and placement of contents copy and readability of text, especially on packages for food, pharmaceuticals, and chemicals

- *claims* regarding benefits of the product inside the package

- *intellectual property issues* covering such critical subjects as trademarks, copyrights, and trade dress

Who is responsible for all of this? Who should sort out what legal requirements apply to the packages being developed?

Eric F. Greenberg, an attorney specializing in packaging law, in his book *Guide to Packaging Law,* says "It is now long past time that a unified body of information called 'packaging law' be defined and identified. Once this is done, packaging professionals can look with more certainty to a single source for study of law, regulations, and concepts they need to know about properly understanding the legal requirements that affect their field."

Meanwhile, marketers, designers, and packaging engineers have no choice but to thread their way through a plethora of federal, state, and local laws and regulations. Design consultants are familiar with many of these, especially those that apply to many packages. But the legal process of applying packaging laws and regulations to the multitude of retail packages is too complex and too specific to fall within the responsibility of a single office. Eric Greenberg advises relying on the expertise of the marketer's legal department or outside legal counsel because "a little knowledge can be a dangerous thing. A nonlawyer's study of packaging law should be the beginning, not the end, of his or her learning."

Since package communications occasionally conflict with packaging laws (as exemplified by the ongoing debates about such subjects as bottle return deposits and what constitutes "light" food), it is especially important to make sure that each package complies with all legal and regulatory requirements, both structurally and graphically, and to help marketers and designers avoid potential trouble.

It is important for the marketer to understand that even if the design consultant is familiar with many of the legal regulations, such as net weight, sizes, Nutrition Facts panels, size requirements for mandatory text, and caution devices required on packages for chemicals, to name just a few, *final responsibility for all structural and copy elements is always that of the marketer.* For that reason, no package should ever be introduced into the marketplace without legal consultation and advice, even if the package underwent only a minor modification. The cardinal rule should always be Better to be safe than sorry!

Packaging Communications or a Legal Document?

More and more a package's format seems to resemble a legal document. The regulators continuously impose new requirements on the marketer, and these laws keep changing. Consequently, while the front panel of the typical package becomes the selling medium, the other panels often look as if they were written by lawyers—and often they are.

In an effort to "protect" the consumer from misunderstandings about a product, particularly in the food, beverage, and health care categories, government lawmakers and regulators periodically make new rules meant to protect the consumer from possible harm or contribute to a better understanding of what he or she is about to buy. One of the best known examples of this is the creation and implementation a few years ago of the Nutrition Facts panels that must appear on nearly all food packages, describing fat, cholesterol, and sodium contents and other health-related information. Similarly, pharmaceutical packaging must list detailed information regarding indications, directions, and warnings. While on larger packages there are few problems accommodating such information, this can become a major problem for the designer on very small packages, such as for many pharmaceutical products and some cosmetics. While exceptions or slight changes in formatting this information on small labels and packages are allowed by the Food and Drug Administration (FDA), many manufacturers' legal departments do not want to take any chances and insist on unequivocal meticulousness. This often results in huge amounts of copy, leaving the designer to accommodate this information in legible condition within the minute proportions of small packaging units.

While Congress and the FDA, which studies and most often sets these policies, have complained that some package labels read like legal contracts, they ignore the fact that *they* are the source for creating many of these situations and that marketers believe that they are following the essential information. More recently, the FDA has proposed simplifying packaging communications on over-the-counter medicines, which will affect the *back or side panels* of packages. *Discard* becomes *throw away*, *consult a physician* will be changed to *see a doctor*, *hole* will replace *perforation*, and other revised language will make it easier for users to understand the benefits and potencies of the products they buy. Some industries, such as pharmaceutical manufacturers, welcome and support these changes.

Language simplification is desirable in all communication media from packaging to advertising, and designers and their clients need to avoid using words that may be confusing to the consumer. While this should be the rule for all package copy development, marketers must be cautious so that consumers clearly understand what they are about to purchase and are not given an opportunity to accuse manufacturers of false or misleading information. For that reason alone, legal counsel should always be sought to assure that *all* copy, whether promotional text, copy describing usage directions, or protective information (such as in connection with pharmaceuticals, chemicals, aerosols, and flammable liquids), is precise and legally supportable, if challenged.

Trademarks and Trade Names

A trademark can be a word, symbol, logo, a combination of words, or even a package shape or color that distinguishes one brand or product from others. Trademarks are a company's most valuable assets, and a great deal of money and time is spent in protecting them. In most companies, trademark protection receives top priority; trademark guidelines are updated regularly and published to ensure internal compliance.

Trademark protection is relatively straightforward. In the United States, a company can own a mark by simply using it in commerce or on products or by filing an application with the U.S. Patent and Trademark Office stating its *intent* to use it. U.S. trademark registrations have an initial term of only ten years, but they can be renewed in perpetuity for as long as the mark is in use. However, both names and trademarks should be run through a search procedure prior to adoption and use to make sure that no other company has a brand name or trademark that is sufficiently similar to conflict with the proposed design.

Brand name and trademark consultants, as well as designers and marketers themselves, can access various reference sources. This is best accomplished in incremental steps.

The first step, after developing a number of alternate brand names, trademarks, logos, or package designs, is to conduct a *preliminary* clearance search to separate all available candidates. While this is only a *preliminary* search, that is, a quickly executed informational overview (usually requiring between one and three weeks, depending on the extent of the search), and is not comprehensive enough to be legally conclusive, this procedure will separate the more obviously nonavailable or questionable candidates from those that are feasible. Because of its preliminary nature and swiftness of execution, the costs for a *preliminary* clearance search are moderate.

Many brand-name consultants and marketers themselves have the capacity to check the availability of trademarks and brand names through Internet or federal and state databases. In addition, trademark attorneys regularly perform these searches, and consulting services specialize exclusively in brand-name and trademark research.

The second step is to probe more deeply into the availability of the selected name candidates. While the *preliminary* search is intended to peel away those marks, logos, or package designs that conflict with others or are similar enough to be confused with others, a further *common-law* search will determine in greater detail whether the remaining candidates are really free to use or whether they are questionable and may have to be negotiated with the owners of similar names, marks, and devices. While this search can be done by a marketer or a consultant, it is advisable, at this stage of development, to enlist the marketer's legal department or an outside trademark attorney.

It should be kept in mind that trademark laws that apply in the United States are not applicable in European countries or in Asia. What may not be confusing identification in one country may be confusing in another. For example, while in the United States trademark rights are acquired by adopting a mark and using it on products that are sold in commerce, in many foreign countries trademark rights in a name or brand can be obtained by merely registering it with the applicable governmental authorities. This can lead to situations where foreigners acquire the trademark rights of a U.S. mark and then require the U.S. owner to buy back those rights. For that reason, U.S. companies who market their products abroad should always protect their marks and brand names by registering them properly and promptly.

International recognition of trademarks and other identification devices can be achieved through registration in Europe on the International Register (also known as the Madrid Agreement or the Madrid Protocol) or with the European Community Trade Mark Register. To do so, proper legal counsel is virtually mandatory.

Identifying ownership of a brand is often fraught with legal difficulties. Nonuse of a registered trademark for a period of several years, for example, may lead to loss of trademark rights by the original owner. Also, if the original owner of a trademark allows others to use it without authorization, the mark may eventually become generic, as happened to *aspirin*. A typical example is the Ralston Purina Company, which marketed one of their products under the brand name Cat Chow for many years but almost lost the exclusive rights to this name because the words *cat* and *chow* were challenged by competitors as being generic. By adding Purina in front of the name (that is, Purina Cat Chow) and displaying the name on the packages in a distinct lettering style, the company was able to protect this valuable property as a trademark. These examples amplify the importance for trademark proprietors to be vigilant in protecting their trademarks from unauthorized use.

Trade Dress

Trade dress, however, is a murkier area for designers and marketers and one that is receiving considerable attention as marketers try to establish meaningful differences in the competitive environment. Trade dress describes the *overall* look of a package—its graphics, colors, shape, and other elements that contribute to a distinctive identity.

Unlike a trademark, trade dress is difficult to register. However, *specific components* of the design are protectable. Trade dress is sometimes copied to confuse users into thinking that they are purchasing the real thing, the original product, while they are really being duped into buying a product "knockoff." The term *knockoff* has become part of our lexicon. Once limited primarily to private labels that copy the look of nationally advertised brands, copycat package design is more prevalent than ever as marketers attempt to imitate the leaders of product categories.

All elements on the package—name, logo, color scheme, illustrations or photographs, even the shape of the container—are considered part of the trade dress. While common sense dictates that copying someone else's brand name, logo, or trade dress is an invitation to a lawsuit and damaging to a company's image, sometimes marketers and designers can get too close to a competitive packaging format, thereby risking challenges that may result in serious and costly legal action. Situations such as these, deliberate or accidental, are easily avoided through awareness of what goes on in the product category and research in trademark registers. The computer has greatly facilitated trademark and trade dress research, and the advice of the marketer's legal department or outside legal counsel should be heeded.

Although a company can sue to protect trade dress, these are delicate and difficult issues for the courts to decide since each situation is different and the degree of imitation and the intent of the imitator must be evaluated. Consumer products giant Procter & Gamble has filed numerous trade dress infringement lawsuits against other marketers, even its own customers—drug chains and supermarkets—for knocking off packages of well-known Procter & Gamble products such as Pantene and Head & Shoulders shampoo, Secret and Sure deodorant, and others. Because of its clout as a major product manufacturer throughout the world, Procter & Gamble usually wins.

However, such lawsuits are not always successful. Recently, a federal appeals court dealt a serious setback to Chesebrough-Pond's when it claimed that a retail chain, Venture (one of its customers), was selling its Skin Care Lotion hand cream in packages that featured the same shape and label colors as Vaseline Intensive Care skin lotion, one of Chesebrough-Pond's top brands. The court, however, decided that since Venture displayed its logo in a prominent way on the label, consumers would not be fooled because they were familiar with the Venture logo on signs, trucks, advertising, and other packages. This opens the door for other private labels to create a "knockoff" and get away with it while challenging Chesebrough-Pond's to go through the expensive process of developing new proprietary bottle structures, label graphics, or both.

In some situations a marketer may be willing to deal with accusations of trademark and trade dress infringement as a part of doing business. The Coca-Cola Company for example does not tolerate any attempt by others to usurp or weaken the competitive advantages of its brands through trademark infringement. Witness Coca-Cola's successful battle with Sainsbury in the United Kingdom, which resulted in Sainsbury's having to change their cola packages to reduce their resemblance to Coca-Cola's. But in 1997, Coca-Cola was challenged by three small marketers when it introduced Surge, a citrus-flavored soft drink. Two of the litigants claimed ownership of the Surge name for beverage-related products, while the third claimed that it had previously been using a slogan similar to Coca-Cola's "Feed the rush." Certainly, Coca-Cola's legal department must have been aware of these potential conflicts; apparently the value of the Surge name was considered more important than the cost of fighting legal action.

Color My Image

In the ongoing and intense search to carve out a competitive advantage for their products, companies often try to establish recognition for a brand through the use of color or a combination of colors on packaging that will become readily identifiable with the product. Kodak's yellow, Fuji's green, Coke's red, Hershey's dark brown, and the green of Bayer aspirin's worldwide packaging are but a few examples of the value of strengthening marketplace positions through color. Other brands were not as careful or made no effort to guard the equity of their originally proprietary colors. Prestone's yellow container color is now used by several other antifreeze brands, and Healthy Choice's green packaging became the forerunner for numerous other packages for "healthy" food brands.

In some circumstances, color *can* qualify as a trademark, though this is very difficult to support legally. But a 1989 decision by the U.S. Supreme Court could have significant impact on packaging. In a well-publicized case, Qualitex Company sued Jacobson Products Company for closely imitating the distinctive, trademarked green-gold color of their large line of dry cleaning pads, a well-recognized, leading product used by dry cleaners for more than thirty years. The Supreme Court agreed with Qualitex Company by ruling that *a company can have a particular product color as its registered, protected trademark*. Similar cases, however, have been examined by other courts with varying results.

Color as a valid trademark depends on whether it can be proven that the public associates a particular color with a particular source. However, when a color is *essential to a product's use or purpose*, it cannot be trademarked. For example, if a particular color is used for *all* products in a particular product category (such as the yellow safety icons on chemical labels), such colors cannot become the property of one company.

The bottom line? In *New Shades of Trademarking*, lawyer Eric F. Greenberg suggests, "If you are thinking of claiming trademark rights in a color, choose a color that is unusual, at least in the context in which it is used, and *promote* that color as a trademark, perhaps by emphasizing it in advertising."

All of this is but a small sampling of the legal issues that may have to be addressed in the course of package design development. In addition to those mentioned, there are a number of sources for obtaining more detailed information regarding legal compliance on packaging. For specific information, the *Federal Register*, published five times a week, carries regulatory announcements. The Government Printing Office in Washington, D.C., issues books and reports on a wide range of legal topics, including those relating to packaging. You can reach this office by calling (202) 275-2091 if you are interested in a particular subject.

In addition there are a number of government bodies that can furnish information on packaging issues. Most are located in Washington, D.C., and include

- Consumer Product Safety Commission
- Drug Enforcement Administration
- Environmental Protection Agency
- Federal Trade Commission
- Food and Drug Administration
- U.S. Department of Agriculture
- U.S. Department of Commerce
- U.S. Patent and Trademark Office

There are also numerous other sources that will help you stay on top of developments in packaging law. Business magazines, general press, and industry newsletters offer valuable insights. Memberships in professional organizations will help update you on changing events. And don't pass up your local library as a truly productive resource for information.

Client-Consultant Relationships and the Law

In addition to the variety of legal situations relating to package design, marketers and designers face a number of other legal issues that can impact on the design development program. These can sometimes be complex and need to be addressed with or without the advice of legal counsel. They concern a variety of aspects concerning the client-consultant working relationship, such as confidentiality during and after completion of a project, confidentiality within the design organization and its employees and freelancers, responsibility of designers and clients should the final package be challenged legally, and ownership rights regarding the designer's work.

Package design, as has been pointed out in earlier sections of this book, is a *business*. Therefore, to ensure that the project is mutually rewarding and to minimize the risk of misunderstandings, the relationship between marketer (client) and designer (consultant) should adhere to sound business principles.

For too long, the commercial world viewed designers as artists who understood and cared little about *business* and who worked purely for the love of art. Not to in any way minimize the creative side of the designer's contributions, any designer who shows disdain for financial rewards and efficient business procedures is going to encounter serious problems in client relationships that may adversely affect the quality of the work. The best interests of both client and consultant require *clearly articulated agreements* outlining what is expected of the consultant and what the responsibilities of the marketers are. A contract or written agreement should not be viewed as an adversarial document but as a framework to help ensure a sound and productive working association.

Contracts

There was a time when very successful design consultants, including the authors of this book, did not see any need for signing formal contracts with some of their clients. A handshake or a brief memo of understanding was sufficient. This practice may have been appropriate when the consultants' primary contact was the president or someone else in a key position at the company. At that time, companies typically experienced minimal employee turnover, and consultants could depend on the client team members to be responsible for the entire course of the project. But times have changed. Today, client team members are more likely brand or product managers, and it is not unusual for client personnel to be transferred to another position or leave the company in the middle of an assignment. New people bring with them different expectations and experiences. They may want to make changes in the project or head in a new direction. Service fees and expenses incurred by the previous team may be put into question. For these reasons alone, a written agreement between client and consultant, most often a part of the consultant's proposal, is essential so that misunderstandings about billing fees and other expenses will not ruin an otherwise productive relationship. How design development proposals are structured was discussed in Chapter 5, "Selecting the Designer."

One of the issues that occasionally leads to misunderstandings between the marketer and the design consultant is the manner of handling out-of-pocket expenses. Out-of-pocket expenses are costs incurred by the designer in the course of the design program for external purchases or for subcontracts. Misunderstandings occur primarily because most marketers are more accustomed to their working relationships with ad agencies than with consultants. In the past, advertising agencies historically *marked up* expenses by an agreed-upon percentage. But this has changed. Today, most ad agencies pass through expenses *at cost* to their clients. However, the design firms' considerably tighter budgets cannot as easily absorb additional outside expenses, and thus they often continue to follow the traditional markup method of adding a *carrying charge* (usually 15–20 percent) to their out-of-pocket expenses. Clients are sometimes surprised when the first invoice of a design project arrives showing a carrying charge added to out-of-pocket expenses. Agreements with the consultant should specify, *in advance* and *in writing,* all matters pertaining to costs and billing procedures.

Ownership of Designs

Designers follow a creative process that starts with a broad exploration of alternatives and directions that, after screening and evaluation, is narrowed down to a manageable number of final candidates that are developed in greater detail, as described in earlier chapters of this book. The initial selection of design concepts that warrant further development starts in the consultant's own office but eventually includes the marketer, whose strategic objectives will guide the selection of viable concepts. During this selection process, the marketer is, of course, exposed to a number of preliminary concept explorations. A question then arises: who owns the preliminary concept explorations? Does the marketer own *everything* that was part of the exploratory phase (for which the consultant charges a fee), or does the marketer own only those designs that are selected for actual use (since the contract provides for the design of *one* specific package or packaging line)? This potential source of friction also must be resolved in the agreement.

A particularly difficult ownership issue arises when corporate lawyers want to apply the "work for hire" concept to the working relationship with the consultant. *Work for hire* means that, while working on the marketer's project, the consultant is *in the employ* of the marketer, hence *everything* the consultant's office does in connection with the marketer's project is *the marketer's property*. Clients who are experienced in working with designers understand that if the designer were to relinquish every preliminary idea that is explored—with the implication that no similar concept can be used by the consultant for future assignments, even for packages in entirely different industries—then the designer would be out of business rather quickly. There is not an infinite number of ideas in any creative endeavor. Therefore, the consultant provides design explorations on the basis that the contract provides for ownership of the *end product,* not all preliminary design experimentations, and thus the marketer is entitled to ownership only of the design concept that is finally selected and implemented. Occasionally, an agreement will provide for securing additional design concepts from the preliminary explorations for additional fees.

To reduce overhead and to be more competitive, consultancies occasionally employ independent contractors (freelancers) for specific projects and tasks. This is a perfectly acceptable practice, especially when a specific skill may be required. However, bringing in a freelancer can also complicate ownership issues. When work is created by an employee of the consultant, ownership of the designs automatically belongs to the design firm, who then passes along rights for the final design to the marketer. An independent contractor, however, may claim ownership of his or her designs. This is not acceptable to either the design consultancy or the client. Consequently, marketers should make sure that the consultancy has a contractual agreement with the freelancer that provides for a clear understanding that any work developed by him or her falls under the same contractual obligations that apply to every employee of the consultant.

Confidentiality and Noncompete Agreements

Designers receive from their clients proprietary information. The client expects that this information will not be revealed to parties other than those involved in the project. It is in everyone's best interest that such understanding be spelled out in writing, either as part of the agreement or as a separate document, a Confidentiality Agreement. To further protect their interests and those of their clients, it is essential that consultancies include a confidentiality clause in their employment contracts and agreements with freelancers.

At the time of the client-consultant working relationship, the client usually expects the consultants not to engage in any activities for a competitor of the client for a specific amount of time, usually six months to a year, after completion of the project. This is an accepted practice in advertising and applies to designer consultants as well. Some marketers, however, want to obligate the design consultants to restrict their services not just to products directly competitive to the project at hand but to *all* product categories in which the company is involved. This should be negotiable, since the designer is being asked to relinquish potentially major new business. Some agreements build in compensation for the consultancy to make up for such losses.

Post-Project Legal Responsibilities

Who is responsible if after a project has been completed and the new package introduced in the market, the design is challenged and a lawsuit claiming trade dress infringement is filed by another company or individual?

Some marketers demand that the design consultant accept liability resulting from *any claims* relating to the design project developed by the consultant. This is often specified in contracts prepared by the marketer's legal department or outside lawyers under the provisions of the so-called *indemnification clause*. This clause seeks to shift the responsibility for litigation costs entirely to the package design consultant, even when such claims are frivolous and unwarranted. Consultants who refuse to sign such an agreement risk loss of the assignment.

From the designer's position, since any package design development program is not just the design consultant's creation but has been developed in close coordination with the marketer's project team, this is an unreasonable demand. If the team did its homework during the project and the client's attorneys signed off on the selected design, the consultant should not be held solely responsible for subsequent, possibly fraudulent or unwarranted claims. For this reason, the indemnification clause is invariably challenged by the consultant's legal counsel, who will suggest modifications to the indemnification clause by distributing project responsibility in an appropriate manner. Since marketers usually have substantially greater resources than their design consultancies, along with the added advantage of in-house legal expertise, the consultant considers it fair and appropriate for the marketer to defend any legal challenges to the design solution, unless it is *clearly established* that the consultant was at fault.

Though the emergence of specially tailored insurance coverage for package designers is a promising solution to this dilemma, many insurance companies, concerned about having to defend indiscriminate and often unwarranted claims by opportunistic plaintiffs, shy away from issuing design liability insurance policies. Thus, in the event of litigation, the consultant could be forced to support substantial legal expenses that, under extreme conditions, could drive the consultant into bankruptcy without cause. Hence the consultant's resistance to a one-sided indemnification clause.

A Final Thought Regarding Packaging and the Law

Everyone involved in planning and creating a package knows that labeling regulations seem to undergo continuous change; rulings by the courts regarding ownership of trademarks, trade dress, names, colors, structural design, and shapes seem to create new problems while resolving old ones. Company attorneys and outside lawyers play increasingly important roles in bringing a package to market, and their influence will continue to grow as marketers battle for the same pieces of the pie and as cross-border and global competition generate even more problems.

Remember, the more you know about packaging law and legal requirements and the clearer your agreements are with design consultants, the better and more efficient a job you and the designer are able to do. When in doubt, or if you are faced with a problem that you cannot resolve, consult with the appropriate attorney. You may not always be pleased with the attorney's advice (which may tell you that you are on the wrong road), but you will avoid unnecessary problems.

13 Staying on Target Long-Term

Your company has just introduced the new packaging. The payoff for all your efforts and those of your colleagues and design consultants dominates the category on the shelves. Senior management is impressed; congratulatory messages are filtering down to brand management. The advertising agency is making the new packaging the focal point in print and on TV.

All this is no coincidence. You followed all the rules and did all the right things. You did consumer pre-design research to discover the strengths and weaknesses of your packaging as seen by the consumers, did category audits to uncover weaknesses in competitive packaging and merchandising, set package design criteria that focused on achieving your objectives, selected and retained the most capable design counsel, shepherded the project through the organization, and brought in production and manufacturing staff early in the process to make sure that the packages would fulfill all the technical and production requirements. You have investigated opportunities for offering your products in unique new packaging structures that provide convenience features that your competitors' packages do not have; you tested the final package designs through research to assure you and your management that the new or redesigned packages will be favorably accepted by the consumers. In short, you have done everything that was required to deal with the myriad responsibilities and hurdles inherent in any design program.

Everything worked out the way you planned. Initial reactions from consumers and the trade are encouraging. Your products in the new packages are flying off the shelves. Now that the new packages are fulfilling their objective of selling your products, you can finally relax and bask in the sunshine of a job well done, right?

Wrong! This is no time to relax. Your customers were not the only ones who noticed your new packaging. Your competitors did, too! They are *not* relaxing—they have already begun to evaluate your new packaging. They are dissecting the graphics and any structural innovations you may have introduced. They are wasting no time to strategize a counterattack in ways that will, from their point of view, dilute the effectiveness of your program. This is *no time* to be at ease. On the contrary, introducing new packaging may give you an initial advantage over your competition, but it will also draw your competitor's attention to their own package design needs and accelerate their determination to make your success short-lived.

If new or redesigned packaging is to operate as a meaningful and long-term marketing tool by reinforcing awareness and recognition among current and new users, it is essential that marketers do everything to prepare themselves for the inevitable counterattacks by their competitors from the very moment they launch the new packages. This means that marketers must start auditing the category as soon as the new packages appear on the shelves and *continue* to monitor their effectiveness to ensure that the original goals of the design program are implemented long-term.

To maintain the momentum of the initial packaging introduction, the actions discussed in this chapter should be considered.

Retail Cross-Checking

As an initial step, at least a half-dozen retail stores in which the products are sold should be audited immediately after launching the new designs to observe firsthand how the new packages look under actual retail conditions and whether they really live up to expectation. This should be done when the shelves are fully stocked and again during the course of the day or within a few days after customers have handled and moved the packages around. Get the design consultant involved to help you in identifying key issues relating to the new designs.

- Take photographs of the new packages, as well as competitive packages, in at least one or two typical retail environments, simulating the shopper's experience—as the shopper approaches the packages and first notices them from a distance at an angle and again close-up as the shopper faces the packages on the shelves.

- Observe the shoppers' reactions to the new packages. Do they seem to notice them? Do they stop to inspect them? Do they pick them up for a closer look? Do they put them into their basket or return them to the shelf? Do they pick up a competitive product instead?

- Speak to the shoppers. Do not identify yourself as a representative of any of the products. If asked, just say that you are doing consumer research. Ask questions that will provide worthwhile clues as to the effectiveness of the new package design. Has the shopper *noticed* any change in the new packages? Does the shopper appear to be concerned that the *product* has changed along with the new package design or even that it is a *different* product from what he or she used to buy? Which package in the category is easiest to find? Which is easiest to handle? Which package— yours or the competitors'—explains the product's benefits most clearly?

- And although it is usually dangerous to place consumers in the role of being packaging experts, probe how the shopper rates packages in the overall category. While these informal interviews should never be viewed as a definite study, they can put the marketer into the shoes of the shopper in order to get an overall feel for the shopper's attitude toward the category in general and your and your competitors' packages specifically.

- Store managers and their assistants can also contribute to a better understanding of reactions to the new packaging, both positive and negative. Their opinions are valuable. Do they notice any effect on sales since the new packages have been introduced? Do the new designs present any problems in stacking and displaying? Have they heard any comments from shoppers?

After being satisfied that sufficient time and effort have been extended to obtain meaningful information about the effectiveness of the new packages and whether they have achieved the intended objectives, a report should be prepared that, along with the retail photographs, provides a realistic representation of the situation in the marketplace and establishes guidelines to keep your packaging continuously effective and updated.

Staying on Target

Now that the new packaging has been introduced and is being monitored by the marketer's staff or by outside consultants, it remains to be seen how the new packages can *retain* their effectiveness. Assuming that the initial introduction was successful, the marketer's concern should focus on maintaining the strategic advantages that the new packages provided and whether or not they will continue to serve the products well.

Consumer research, as discussed in earlier chapters, can be used as a methodology for eliciting information from consumers even after the design process has been completed and the packages are in the stores. Just as research served the purpose of defining consumer attitudes to help in the design process, focus groups or in-store interviews can assist marketers to *stay* focused. If unexpected shortcomings in the new designs are detected during the post-introduction monitoring process, marketers can evaluate the need for fine-tuning the packages, such as emphasizing certain information to help shoppers in product selection. This may not require substantial design alterations, but minor modifications of text or design elements can often make a difference in shoppers' acceptance and can usually be accommodated during the next package production run.

Continuous monitoring of the packages through research will often uncover important issues. Most shoppers will not hesitate to express their likes and dislikes of their shopping experience. If the new packages are a redesign of previously existing packages, frame the questions to learn if respondents react differently to the previous packages than they do to the redesigned packages. If they do articulate concerns, such as perceived differences in product efficacy or taste, dig more deeply to determine in what way the packaging is influencing their opinions. Probe for similar concerns with competitive packaging. Use product-related questioning to identify perceived weaknesses in the competitive packages that can then be exploited through marketing and merchandising.

To be effective, this type of monitoring should be conducted on a periodic schedule, cross-checking every three to six months, never waiting longer than a year. Like death and taxes, change in the marketplace is inevitable. Not only will the market change as competitors challenge others by introducing their own packaging innovations, but new brands will enter the arena and the ever present tug of war between major brands and private label clones puts ongoing pressure on marketers to stay one step ahead of competition. This applies particularly if the marketer's product is not the market leader. Staying current with market conditions is a *must* for all marketers of retail products and a never-ending task to *stay on target.*

Establishing Internal Monitoring Controls

There are two main ingredients that will keep a package design program on track: people and practices.

As managers are promoted to other positions or leave for another company, attitudes towards packages that were developed by one group of marketing managers may not be echoed by incoming managers. Newly appointed brand or product managers should not be expected to have the same intense feelings toward the packages that their predecessors did. What is the right way to avoid potential conflicts and rash decisions?

Each company operates differently and will respond to internal attitudes and approaches in a different way. Some companies follow strict rules about packaging and packaging changes; others give the managers more leeway. But there are guidelines that can help marketers make decisions about package design when given the responsibility for marketing existing brands and product lines.

- In most cases, sound principles of marketing suggest that packaging is a marketing tool that should be handled with great caution. Incoming managers should resist the temptation of changing packages to give them their personal imprint. As a general rule, packages should be redesigned only when there is strong evidence of critical changes in market conditions, changes in consumer lifestyles, environmental needs, obviously outdated graphics, or the need for packaging cost reduction.

- Package design changes come in many forms. Some conditions require relatively minor modifications, such as changes in promotional text, changes in legal, regulatory and ingredient copy, product design modifications, or the addition of another product form or flavor. Such changes can be accommodated, in most instances, in ways that will not alter the basic package design and that require minimal decision making. But even then caution should govern all design change decisions.

    Sometimes even a minor change is noticed and can upset the consumer. When Colombo considered package redesign in connection with segmenting changes in its line of yogurt, modifications to update the Colombo logo were also explored. The logo included a small red berry that was thought to mislead consumers to think it was flavor specific. But when the concepts were researched, including Colombo logos that were identical to the existing logo except for deleting the berry, current users immediately spotted the change and objected to it. The final logo retained the red berry and serves as an example of the extreme caution required in modifying even presumably minor package design details.

- Continuous monitoring of the category, as described earlier in this chapter, will give marketers up-to-date information for evaluating the need, the extent, and the timing of package redesign. Packaging design control manuals, discussed in detail in Chapter 7, "Creative Development," can be helpful in maintaining the integrity of packaging elements or, if changes are initiated, record these changes and continue to serve as a guide to subsequent design implementation.

All this emphasis on caution is not meant to imply that dramatic packaging innovations should be avoided. The marketing of retail products would not be served by being overly conservative. The inspiration and excitement accompanying major package redesign is the hallmark of leading designers, whose constant effort in striving for unique new packaging structures and attractive graphics should not be undervalued. The dramatic improvements in package design over the past few years are, to a great extent, responsible for the ever increasing importance of this marketing medium. Despite the traditionally conservative nature of the packaging industry, the success of exciting new packaging structures and daring graphics have shown the way to many new and unique design directions. The packages for wines, cosmetics, and electronics and the packaging of entire brands such as Healthy Choice, which swam against the stream (green food packages were once considered a no-no) are but a few examples of dramatic design innovation that resulted in market leadership.

No matter what position you favor—the conservative approach of rarely making package design changes or the proactive position of seeking uniqueness by experimenting with innovations—the one thing both of these have in common is the need for continuous monitoring of conditions in the product category in order to obtain the basis for making intelligent package design decisions. Thus, when the new product has been launched or package redesign introduced, there is no time to sit back believing that the package design process has now ended. Because for packaging to remain a successful marketing tool, *staying on top long-range* is essential.

14 Where Do We Go from Here?

Looking into the future can be fun and exciting. With the new millennium just around the corner, we are tempted to prophesy all manner of seemingly fantastic ideas about our lifestyles and the world around us in the year 2000 and after. How will all this affect marketing? And if it will affect marketing, what impact will any changes have on packaging and package design?

There seems to be little doubt that lifestyles will change no less dramatically in the next millennium than they did during the past one. And whenever new lifestyles evolve, changes in marketing concepts are never far behind.

How will these changes affect the strength and security of your brand? With constant changes in retailing, will branding be able to maintain its critical role in the marketing of retail products? With the inroads made by the electronic media, will advertising and sales promotion by the traditional means of television and print continue to build and sustain strong brand imagery? Will brands be able to maintain their strength in the face of home shopping, interactive CD-ROM, and pay-per-view communications? Will the electronic media *help* in building brands and brand loyalty? Or will the computer's capability of providing rapid-fire changes contribute to *fractionating* the identity of mature brands? How will packages look in the electronic age? Will packaging play the same role it has traditionally played in translating brand equity into brand loyalty and final sales? These are tough and vital questions that marketers *must* start to ponder as the year 2000 approaches.

It's a well-known fact that consumers are not emotionally wedded to brands as their parents were. In the perplexing array of changes in the retail environment and in communications, brands that are strong today will have to guard their equities or find their vitality subsiding almost overnight and their enviable equities fading from the memories of consumers. Does anyone remember White Cloud? Gleem? Brylcream? Piel's? Once category leaders, these brands have faded even from the memories of consumers old enough to have been users of these products.

What will the role of packaging be in the face of these events, and how will packaging adjust to 2000?

Packaging and the Retail Environment

As the world is changing all around us and as product marketing will inevitably be modified by new retailing techniques and technological revolutions, the need for *maintaining* and *building* on the equity of our brands will challenge the strategic skills and imagination of the marketing community.

Recognizing this, retailers are coming to grips with the fact that in order to survive the electronic onslaught they must strike back with new concepts that will lure consumers away from TV and computers and into their stores. In the 1990s this was achieved primarily by the mega-retailers; *big* equaled *best*.

Lost, seemingly forever, in the mélange of stripped-down merchandising techniques by the mega-merchandisers is the wonderful aroma of bread being baked on the premises, the pungent odor of fresh fish, the sweet smell of fresh-cut flowers, and the splendorous color palette of fresh fruit and vegetables. But the trend among retailers will be to rethink their merchandising strategies in the new millennium. Bored by the purified atmosphere of the mega-stores, many consumers are beginning to question their eagerness to trade minimal service and ambience for low cost. They will become more demanding, less willing to compromise solely for the sake of price. Young, active consumers will seek to *get away* from the tedium of TV and computer screens at home and in business and come to the stores to look for *adventure*. This will apply equally to supermarket chains, mass merchandisers, and especially department stores.

"Twenty years ago," wrote Paul Goldberger in the *New York Times Magazine* in *"The Store Strikes Back,"* "designers made clothes, shoe companies made shoes, perfumeries made scents. And all those things were sold through a one-stop, somewhat generic department store. But retail has become complicated. Consumers have grown both sophisticated and fickle. They want good service, but they also want . . . good value." Thus, department stores are rethinking their strategies and supermarkets are increasing their emphasis on quality products and how to display them to lure the consumer away from TV and the warehouse stores.

Mega-merchandiser Sears Roebuck is dispersing some of their well-known brand lines, such as Craftsman hardware and Kenmore appliances. Instead of one-stop department stores, Sears Roebuck is spawning specialty stores that range from about twenty thousand square feet down to mom-and-pop-sized seven-thousand- to eight-thousand-square feet units with names like Home Life Furniture and Circle of Beauty. These stores promise ambient shopping conditions and knowledgeable service while maintaining their every-day low price policy and promoting their brands under the Sears Roebuck advertising umbrella.

Other brands take a different tack: Niketown in New York City and elsewhere offers their huge range of sports merchandise and memorabilia in a visually dazzling, high-tech environment that appeals to the lifestyles of active, sports-minded, young consumers.

Package design in these stores will play a key function in catering to consumers who either *are* more knowledgeable or who *want to know more* about the products they are about to buy. With minimal sales personnel available in the stores, more than ever packages need to be comprehensible, contemporary, able to communicate clearly, and visually appealing by *harmonizing with their store surroundings*.

To achieve this is complicated by today's audience of *picture*-oriented, often verbally deficient shoppers. For many years sociologists, psychologists, and educators warned that television and its discordant mélange of sights and sounds would produce what is now termed the MTV generation, a stratum of society that responds better to pictures, as offered by TV and the movies, than to the nuances of language. Pictures evoke the emotion in today's young people and move them to action, often without their applying rational thinking. And while computers may not signal the demise of the written word, as some would predict, the verbal is certainly far behind.

For many video-saturated consumers, pictures, symbols, signs, icons, and colors are easier to comprehend than words. The phrase "seeing is believing" will take on increasing and more opportunistic meaning for packaging in the new millennium. Astute marketers, such as Nike, have recognized the power of visual images and have formulated their messages by focusing on the *total* perception of their brand, their products, and their packages by creating a comprehensive, *coordinated* retail environment, such as Niketown.

This new reality presents new challenges and opportunities to retailers—whether supermarkets, mass merchandisers, department stores, or specialty stores—as well as to marketers. In the coming years packaging will no longer be viewed simply as another tool of marketing, separate from advertising, merchandising, sales promotion, public relations, or any other communication medium. For retailers and marketers to compete successfully with the electronic media they will look, more than ever, for more visually *informative* and *entertaining* packaging to play a leading role in enticing the consumer away from TV and computer monitors and to look to *shopping for adventure.*

Packaging in the Electronic Age

While retailers battle the widening influence of on-line shopping, they are also, perhaps paradoxically, *utilizing* the growing range of electronic devices and systems that promise to broaden merchandising opportunities, expedite inventory, and improve operating efficiencies, in addition to catering to the convenience of shoppers.

The introduction of the Universal Product Code (U.P.C.) several years ago was the first step in the computerization of retailing. Fiercely fought against by numerous retailers at the time of introduction, the product coding system is today an indispensable tool for retailers, marketers, and shoppers alike. It should not surprise anyone, therefore, if new electronic systems yet to be devised will flourish and multiply at the retail stores, adding benefits to the shopping experience during the coming years.

One such system, the Portable Shopper, already has made its entrance (see Exhibit 14.1). Introduced by Albert Heijn, the largest grocery chain in the Netherlands, this handheld, simple-to-operate scanning device, resembling a telephone handset, is passed by shoppers over the bar codes of the packages they want

Courtesy of Albert Heijn BV

Exhibit 14.1

to purchase, automatically totaling up all purchases in the scanner. There's no need for waiting in long lines while checkout personnel scan the UPC of each individual package; the Portable Shopper has already done this. All the shopper needs to do is pay for the merchandise at a special express checkout and leave.

It should surprise no one if this leads to even more revolutionary electronically generated conveniences in the next millennium. Visualize the following: the shopper scans a package on the shelf with a personalized, beeper-sized credit device. A duplicate package is automatically dispatched from a separate storage area to a special pickup counter, automatically bagged by a robotics device, and paid for through an ATM-like credit system. At the end of the shopping trip the shopper simply picks up the already bagged and paid-for purchases and departs. Farfetched, you say? Remember that only a few short years ago, the specter of humans circling the earth was only the figment of the imagination of cartoon-strip artists and science-fiction movie producers.

In the meantime, digital communication developers are not sitting on their hands. Taking the cue from how shoppers move around the store, *virtual reality mall surfing* has arrived on the scene. No need to move off the couch, get dressed, and drive to the store. All you need to do is to boot up the computer, touch a few keys, click the mouse and voilà—there is the entire shopping mall right in front of your eyes. You can "walk" up and down the mall looking for your favorite store. Once you find it, you can enter the store, survey the wares, examine each individual package on the shelves, and order what you want, all without ever taking your hand off the mouse. Soon, your purchases will be delivered to your door.

Undoubtedly, this sort of shopping will appeal to on-line aficionados. Mall surfing or other on-line shopping methods could present a substantial challenge to retailers by 2005. Already, a shopping methodology that is gaining ground is causing some grocery retailers to bite their nails: *consumer direct* shopping services that allow customers to order provisions via Internet to be delivered to the shopper's home the same day or bagged for pickup. Customers can select the price and quality of the products from descriptive copy that accompanies the available product list. On-line shoppers can even comparison shop, just as they would if shopping in a store.

When a user is ordering a specific brand of product, one or more competitive brands may appear on screen at the same time, or when a customer is ordering one product, an image of a related product reminds the customer not to forget it. There are currently several Internet services that specialize in computer-generated shopping. Even some grocery chains, instead of bucking the trend, have jumped on the bandwagon with their own on-line ventures.

So, if on-line shopping is here to stay, what will it do to packaging—and package *design* specifically? Will packages need to be designed differently, or will they even disappear as we know them today? There is no doubt that packaging will have to adjust to the conditions that virtual reality shopping will present. But unless distribution methods change dramatically, products will continue to need some sort of container configurations for merchandise that needs to be protected, transported, and stored.

Packaging *graphics*, on the other hand, may change more extensively. Seen on the home computer screen, packages do not appear as clearly, nor can many of them be represented in their actual size. They are also seen peripherally on the monitor, so that information that is currently distributed over several sides of the package or label must be communicated to the on-line shopper in some other manner, possibly by digitally scanning the product code into the computer. Apart from informational needs, attracting the shopper's attention and interest requires such packaging to be *visually* attractive, though probably in a modified manner. Above all, simplicity will be one of the most urgent requirements for packages that are viewed on a home computer monitor. Marketers and designers need to work together to experiment with the most productive way of communicating digitally with the on-line shopper. It is entirely possible that some design consultancies will specialize in developing package design specifically for on-line merchandising or converting existing packages for on-line transmission.

It is also possible that in the coming years consumers will be able to obtain digitized information about the product inside the package, such as nutrition facts, recipe suggestions, indications and contraindications for drugs, chemicals in cosmetic products, and assembly instructions for mechanical and electronic products, by simply scanning the product code on the package into their home computer. In all of this, the package will be an indispensable component in bridging the gap between marketing and communicating with the consumer.

We can prognosticate further what other package related electronic concepts may be conjured up in the coming years. Today, printed circuits prevent pilferage by emitting a loud noise when a package or garment tag has not been deactivated before it is taken out of the store. Taking this one step further, could circuits printed on packages prevent tampering by sending a signal directly to the store manager's office? And why could not printed circuits on packages provide for the automatic setting of the correct temperature for food in a microwave oven, similar to the codes on film containers that activate the respective ASA settings in cameras? Could printed circuits on tiny labels, such as on many OTC pharmaceutical packages, transfer dosage instructions to a home video or computer screen, thus making the copy more readable for mature consumers?

The electronic age has only just begun, and its effect on package design is as intriguing and challenging as it is inescapable. Whether it will be a blessing or a concern, it behooves marketers and package designers to prepare themselves for an adventurous trip into the spheres of virtual reality.

The Globalization of Packaging

Since the end of World War II and the subsequent rebuilding of industries in Europe and Asia, *global marketing* has been one of the most talked-about subjects in the marketing community. And the package design community has been anxious to ride the tide. Commitments resulting from the creation of the EC, NAFTA, GATT, and other transcontinental agreements have stimulated the explosive growth of global trading. The advances in transportation of both people and goods throughout the world, the coordination of global manufacturing facilities by major corporations, and the proliferation of global communication options have made the world a single, homogeneous entity—or have they?

While the globalization of marketing communications enables marketers to spread messages about brands and products into ever more remote locales, most of these messages will be new and strange-sounding to consumers whose lifestyles and customs may differ widely. The availability of more and more products, combined with more and more competition among marketers, will fuel the debate over how to design packages for global marketing. Should there be commonality, so that the same look is marketed everywhere? Should the look of the package change from region to region or from country to country to accommodate cultural characteristics?

We tend to talk about the differences between Latin America, Europe, Africa, and Asia as if each of these were homogeneous continents, overlooking the fact that there are not only substantial differences in customs, tastes, and shopping habits between these continents but even *within* each of these regions. Brazil and Argentina are divided linguistically, though they are neighbors; Germans and Italians differ ideologically and gastronomically, though both are European; and Japanese have little in common with Indians, even if both are Asian. The fact is that all of these regions are far from being homogeneous. When this translates into global marketing strategy, what type of package design will work effectively in a global context? How will brand names, languages, pictures, and colors fare in the global environment?

There are many examples to demonstrate the traps that await the unsuspecting marketer who attempts worldwide distribution of brands or products without careful orientation to make sure that the words on the packages, including the brand names, and the visuals will stand up in the countries in which the products are to be marketed. Among the annals of experience are some examples that sound humorous but that courted disaster:

- A once popular Chevrolet car model Nova was a laughingstock when introduced in South America, where *no va* translates into "doesn't go."

- The word Omo, representing a leading brand of detergents in Brazil, has homosexual connotations in neighboring Argentina.

- The Vick's brand of cough drops, sold all over the world, is spelled Wick's in Germany, where the pronunciation of the original brand name has a socially unacceptable meaning.

- The number 4 is considered so unlucky in Japan that marketers not only avoid the number but even abstain from packaging four items together.

- In Egypt, green—the national color—must never be used commercially, especially not on throwaway packaging; and the colors black and white can have diametrically opposite meanings in various countries around the globe.

- A U.S. company wanted to celebrate the nations participating in the 1996 Olympics in Atlanta by reproducing their national flags on the company's globally distributed packages. When Saudi Arabia protested because their flag contains words from the holy Koran, the packages had to be hastily withdrawn.

Nor are language, colors, and pictures on globally marketed packaging the only complications. Regulatory requirements in various countries can be a daunting problem. Packages that are acceptable in the United States can run into regulatory restrictions elsewhere. Environmental regulations in some countries are stringent. In Germany, the Green Dot (Grüne Punkt) system requires all packaging to be returned to and collected by the source of sale, from which they are picked up and processed for disposal. Plastics are frowned upon. Other countries have their own, frequently contradictory policies regarding packaging materials, sizes, language specifications, and numerous other regulatory guidelines.

There are also major differences between the marketing philosophies of various countries. For example, one of the major differences between Japanese marketing theory and the marketing mindset of most Western marketers is that in Japan, *corporate identification, long-range planning,* and *new product launches* are of key importance, while in the United States and Europe, return on investment, that is, brand building and equity, are the primary objectives. The effect of these philosophical differences on package design is substantial. U.S. marketers will devote endless hours and large amounts of money to protect the equity of their brands' packaging, reflected in package designs that evolve gradually over the years. Japanese marketers have no hesitation to change package designs frequently and dramatically without any concern for visual continuity.

Package design for worldwide-distributed products must accommodate these divergent factors. The answer to much of this is research, research, research! Marketers need to know their consumers wherever their products are marketed, requiring careful orientation to make sure that the brand names, words, visuals, and even structural features on the packages are appropriate in the market where the products are to be sold. Closures and dispensers, for example, that are greeted enthusiastically in one market may fail miserably in other places. Every detail on the global package must be carefully considered. Cultural differences and customs from country to country and, yes, even *within* some countries can be substantial.

To "think globally and act locally" makes sense. Marketers need to accept that the world is not one homogeneous place and that consumers respond more favorably to marketing and packaging that shows sensitivity to local preferences of shapes, words, pictures, and colors. Although a single brand umbrella and a common packaging format produce significant economies, can a marketer risk alienating audiences in other regions? In the coming millennium, cross-border marketers will encounter more and more of these challenges as they seek to penetrate customer sensibilities around the world.

As the stakes get higher, post-millennium marketing will need to access consumers in a transnational marketplace where the cultural and linguistic differences that complicate the selling process must be translated into appropriate package design. Printed information in more than one language, although appropriate in many situations, will not be sufficient to provide the emotional need for reaching into the shopper's psyche in markets around the globe. Marketers will need to learn much more about the *people* in these markets in order to generate positive emotive reactions to a brand or a product. Few marketers are as fortunate as Coca-Cola, whose bottle shape is synonymous with the product regardless of the words on the package. And even for Coke, the best-recognized brand in the world, the words on the bottle change from country to country.

To respond to these demands, marketers must make sure that their design consultants have the ability, experience, and facilities to handle packaging on a global scale, whether this requires creating a single look for packages distributed throughout the world or whether the brand strategy calls for harmonizing a large number of packaging units that take regional differences and linguistic nuances into consideration. The task of analyzing all this is often not only complicated by linguistic or cultural differences but often influenced by strong feelings of national sensitivities among the participants of the global marketing team. It is not unusual in global design programs that resentment of the authority exerted by the headquarters country over the other team countries gives the development process competitiveness that the design consultant must handle with a great deal of tact.

The ability of the design consultant to understand all this as well as how to create a global brand architecture that provides for the proper positioning of various packages in diverse regions throughout the world is key to achieving a successful global program. There are design consultancies that specialize in services that accommodate global design development. As the need for such services increases, more and more designers will form alliances or mergers with colleagues in other countries, resulting in larger design consultancies that are capable of handling design programs for both global and local markets. Others will affiliate with ad agencies in order to build stronger "one-stop" communication powerhouses.

Whatever the next millennium will bring—dramatic changes in retailing, advances in electronic communications, the globalization of packaging, or any other momentous marketing-related events—one thing that will never change is the significance of package design to market retail products successfully. To help the marketer achieve this objective and make it an enjoyable experience has been the goal of *The Marketer's Guide to Successful Package Design*.

Insights

One of the characteristics of brand identity and package design is that every project creates new challenges and each challenge requires the search for new and unique solutions. What better way to find solutions than by dipping into the rich well of experience of those who previously executed brand identity and package design projects to learn from them what routes can lead to success or failure?

The *insights* on the following pages describe real-life projects. Each describes an actual case history. Only the names of the companies, the identity of the brands, and in some instances, the categories and details about the procedures have been altered in order to safeguard confidentiality. Browsing among the *insights* may spark ideas that apply to your own needs and requirements.

Insight 1: Combining Creative Technology with Innovative Marketing

Background

Torrid Corporation markets several lines of fruit juices and chilled fruit drinks. The company's technical staff developed a proprietary technology that had the potential for manufacturing a unique consumer-friendly plastic container, far superior to anything in the marketplace. The fruit beverage category is dominated by a few well-known brand names, all of which are sold in standard packaging structures. Torrid's marketing management believed that successful adaptation of this new technology to packaging would provide Torrid with a powerful competitive edge and catapult the company into a leadership position. However, none of the mockups developed by the internal technicians evoked strong consumer appeal. Management realized that in order to market the new technology, the solution must be founded on sound marketing principles and consumer needs and preferences, not on technology alone. To investigate opportunities, management screened several product design consultancies and selected one that had earned a superior reputation among beverage marketers.

Challenge

The challenge was to explore how the innovative technological breakthrough could be translated into consumer-friendly packaging and a long-term marketing advantage for Torrid, a solution that could be adapted to *all* the company's juice and fruit drink lines.

Analysis and Procedures

At a pre-design meeting, Torrid's marketing and research representatives thoroughly briefed the consulting firm regarding the proprietary technology and its potential applications to packaging as well as its technical limitations. The competitive situation in the juice and fruit drink markets and the company's strategic plans were discussed. It was decided to use the *chilled fruit drink line* for the structural design investigation.

The project team, headed by the design consultancy's creative director and the client brand manager, visited representative retail outlets to survey the fruit drink category, as well as other beverage categories, to obtain additional insights into packaging practices.

Parameters and schedules for the project were clearly defined and design criteria delineated. In addition to communicating the value-added benefits that the new technology offered, structural packaging criteria encompassed a number of issues relating to manufacture, retailing, consumer usage, and aesthetics, including

- credible shape for a beverage product
- easy to open, dispense from, and store at home and in the refrigerator
- convenient to stack on retail shelves
- suitable for manufacture in PET
- right size and shape for portability
- tamper resistant
- recyclable
- easy to fill in production
- good shelf life; not affected by light
- suitable for labeling and brand identity

Based on the technology and design criteria, the designers developed a range of structural concepts, including shapes, closures, spouts, and other features. These concepts were presented as two-dimensional drawings, applied to single-serving and twenty-ounce containers. The client and consultants selected three concepts that were further developed—features refined, colors and other details added—and presented to senior management in acrylic model form to better evaluate the unique shapes that the technology made possible.

At the presentation two concepts were chosen for further detailing. Working drawings were prepared and translated into vacuum-formed functional models, each in three sizes. Preliminary graphics were designed to bring brand reality to these structures.

In focus group interviews conducted among adults and children, *both* concepts scored high with the respondents. While this was flattering, it made the final selection more difficult. After further discussion, one concept was chosen by marketing management and the design consultant because its shape was considered more unique and offered a larger area for brand graphics and product description.

Working closely with the design consultancy, Torrid's engineers developed package production specifications for the entire chilled fruit drink line. Label graphics were finalized by the design consultants.

The advertising agency, which had been brought into the project during the initial stages, created a series of TV spots that highlighted the innovative packaging concepts and consumer benefits. A major promotion campaign aimed at young people was launched, and a nationwide sampling program further supported the product launch.

Results

The technologically advanced packaging instantly leap-frogged Torrid from relative obscurity to the number-three position in the chilled fruit drink category in the United States. A version of the innovative packaging is being developed for the company's other juice products and for new products now in the pipeline. Overseas introduction is planned. The success of this venture illustrates how a marketer's willingness to invest in a concept that looked promising lead to innovative marketing by marrying creative technology with imaginative, consumer-oriented structural and graphic design.

Insight 2: Improving Consumer Perception of a Store Chain

Background

Johnson Claiborne, one of the leading department store and mass merchandising chains, also is among the top marketers of household paints; its stores sell hundreds of millions of dollars in interior and exterior paints and accessories such as brushes, rollers, spray-paint equipment, and protective cloths. The paints, displayed according to price points, range from relatively low prices to premium quality that is comparable to nationally advertised brands.

As customer self-service becomes the norm and knowledgeable sales personnel vanish, packaging takes on increased importance as the primary way of explaining end-use benefits and product values to the shopper. Paint is an especially difficult category for consumers to understand. Younger people bring minimal experience to the situation. More women than ever work at home improvements, and painting is a mystery to them, as it is to many men. Shoppers find the mass array of paint labels confusing and intimidating. Why is a $9.95 interior paint less efficacious than a $16.95 interior paint? Why is one brush preferable to another? How much paint do I need for my job? Such questions often drive the consumer to the *local* paint store, where a helpful clerk can provide some answers even if this means paying more for a well-known brand.

How then can the consumer be induced to buy one of the higher-priced items in the *chain's* line? While these paints are advertised to be comparable to the established advertised brands, most consumers do not expect to pay the same price for the chain's product as for an established brand with which they are familiar. This applies even more to high-profit accessories. Shoppers may buy paint at the chain but they go elsewhere for brushes and accessories that they perceive as being superior to the chain's products.

Challenge

The challenge was to improve the consumer's quality perception of the chain's paints and accessories by developing a packaging and merchandising system that would prove consumer-friendly for *all* products in the chain's paint department and motivate the shopper to select higher-priced accessories and collaterals along with the chain's paints as part of the painting experience.

Analysis and Procedures

Visits to four representative paint departments in each of three geographic regions, on-site study of the ways in which the products are displayed and sold, and informal interviews with sixty consumers in the stores confirmed that the lack of understanding not only resulted in lost sales of paints and related items but also negatively affected the consumer's image of the chain's other departments. Shoppers said that if the chain couldn't understand paint, how could they trust the chain to explain hardware, gardening, household appliances, and other product categories? Some shoppers questioned the store's motives in displaying so many price points, wondering if this was an attempt at causing confusion that would encourage the consumer to purchase the higher-priced items for security.

Since no other major retailing chain had addressed this problem successfully, the consultants saw an opportunity to help their client establish a leadership position in "caring about the consumer," which would also enhance sales in other departments.

Criteria that clearly articulated objectives and project parameters were formulated with the chain's marketing executives. Since several different groups are involved in any packaging or merchandising project for a retailing chain, it was critical that everyone be tuned in at the outset to assure that the solutions would be implemented properly.

The consultants approached the situation in its totality by developing a brand architecture that divided the product line into two segments—*interior* paints and accessories and *exterior* paints and accessories. Product descriptions were written for each price point, resulting in the elimination of three of the seven original price points for paints, thus simplifying the merchandising mix. Graphics for the lower-priced items were simple and more generic-looking, while the designs for the higher-quality products were created to compete with nationally advertised brands. Regardless of price point, *all* the labels were easy to understand and explained product benefits and user advantages. Using the chain's name and fine reputation as *endorsements,* the premium paints (interior and exterior) were given brand names and supported by national advertising.

Packaging graphics for brushes and other accessories were linked to the paint can labels so shoppers would look for items that appeared to be specific to the paints they were buying. A premium line of accessories was made more distinctive through unique structural packaging developed for brushes, rollers, and other items.

With packaging as the major component of the merchandising program, department signage, displays, and other materials were coordinated to truly educate the consumer and bring about a satisfied and more knowledgeable shopper. New, informative signage was developed to identify the paint department, to direct consumers to the products in which they were interested, and to explain the differences in paints at the shelf level. The overall department environment now was user friendly and clearly identified all products available in the department.

To make the shopping experience even more pleasurable, a simple computer system was installed in the department to enable shoppers to press buttons depending on their needs and to educate them as to the kinds of paints and accessories they would need as well as the alternatives available to them.

Instead of the chain's usual practice of using central checkout counters for all merchandise, an integrated paint department checkout counter was recommended by the consultant, which would allow for display of high-profit impulse items and make the shopper feel that he or she was in a paint store rather than a huge retailing environment.

Results

The paint shop concept was tested in three of the chain's stores. Sales in these stores increased from 28 to 35 percent. Following this successful launch, the new system was installed in all the chain's stores in the United States.

Insight 3: Integrating the Identity and Packaging of Two Merged Corporations

Background

Easter and Navo, two giant industrial holding companies, merged to form Eastvo with combined sales of $30 billion and in excess of 250 products that are marketed to businesses and consumers in the United States and overseas. These products range from aerospace and plastics to truck tires and motor oils. Following the merger, an identity system was developed to communicate a consistent image for the new company name across all product lines and address key publics such as the financial community, media, the trade community, customers, suppliers, employees, government agencies, and the general public.

Any successful business combination not only retains but enhances equities in brand and product names that have been built and nurtured over the years. Unlike some situations in which management determines that the corporate name dominates communications and marketing, in Eastvo's case the decision was made to link identities between the parent and the brands, so each could support the other. The design consultancy developed an identity system for corporate and operating unit administrative materials, plant and office signage, vehicles, and advertising signatures and then addressed the complex issue of packaging. With many existing brand names and new ones on the way, management recognized that a system needed to be adopted that allowed the corporation to go to market with "one face," yet not weaken established and profitable brand recognition.

Challenge

The challenge was to create a packaging system and guidelines for implementation that would integrate the corporate identity with each brand's identity, support product marketing, strengthen synergistic communications and marketing benefits throughout the corporation, and extend awareness of Eastvo as a technology-driven organization. The variety of packaging structures, ranging from cylinders to cartons to labels in dozens of shapes and sizes, complicated the problem. Any new design format must accommodate each product's requirements. Although corporate management understood the importance of the packaging program, it was critical to elicit the support and cooperation of the operating units.

Analysis and Procedures

During the corporate identity program development, which the design consultancy created following the merger, a liaison was appointed in each operating unit to work with the corporate communications office on implementation issues. These operations' representatives were now advised that a company-wide packaging program was being developed.

To start the package design program, the design consultancy and Eastvo's corporate identity manager met with the marketing and purchasing staff at each business unit to learn about their packaging practices and any special requirements. The benefits of adopting a universal packaging system were explained and questions answered. Since the corporate identity system had already been implemented on other materials throughout the organization, the business units were prepared for packaging changes.

Representative packages from each business unit were collected and studied for copy, color, and brand identity usage. Special attention was paid to possible equities established by certain brands that might need to be integrated in some manner.

To show corporate management the image and communications consequences of different package design formats using varying degrees of corporate and brand emphasis, three alternate brand architectures and the resulting design formats were presented in the context of package prototypes:

- primary emphasis on corporate identity on the main panel

- primary emphasis on brand identity with minimal corporate identity (relegating corporate identity to side or back panels)

- shared emphasis of corporate and brand identities

The first two options were ruled out as inconsistent with corporate strategic thinking, putting too much emphasis on one identification at the expense of the other. The third option was more on target but suggested possible visual conflicts between the corporate and brand identity. It was decided to further explore how to achieve an appropriate balance.

After exploring a number of alternatives, the recommended package design format, which was approved by management, apportioned one-quarter to one-third of packages' main panel to the corporate logo, depending on the size and shape of the package unit, and the remaining three-quarters to two-thirds of the area to brand identity and product description. This solution afforded suitable visibility to the parent company without diluting impact for the brand. The format included consistent styling, color, and positioning of the corporate logo in the upper section of all the packages. To avoid visual confusion and for a contemporary and clean look, brand logotypes were replaced by the corporation's typographic styles, except when longstanding equities dictated retaining a brand logotype.

Results

Eastvo and its business units now benefit synergistically from an identity and packaging strategy that maximizes the recognition and marketing equities in the corporate and brand names. Sensitivity to change is inherent in any newly merged or acquired company. Eastvo's management recognized these realities and opted to communicate openly with the business units, particularly when any identity changes were contemplated. This approach, which paved the way for smooth introduction of the new corporate name and identity system, worked in the same productive way for packaging.

To further accommodate the smooth integration of the packaging formats, the design consultancy developed standards in written and electronic form, which were distributed to operating unit personnel at several seminars throughout the country and overseas. As new products are introduced and more businesses acquired, these standards are communicated to the appropriate personnel and are expected to be followed. Compliance is supervised by the Eastvo corporate identity manager, who together with the design consultants, is available to assist the business units when needed. Any exceptions to the identity standards must be cleared with the corporate identity manager in advance.

Inventories had been carefully monitored during the package design program's design, which helped make the changeover efficient and cost effective. In this way, the phasing in of the worldwide packaging system was accomplished within a six-month time frame.

Insight 4: Creating and Naming a New Brand

Background

The Criterion Corporation, a global marketer of consumer products that range from wearing apparel to baked goods, has a successful track record as a marketer of a number of well-known branded products that are not publicly identified with the parent. Each of these brands has been marketed on its own, supported by relatively large advertising and merchandising budgets. Because of Criterion's size and clout, retailers readily stock and promote its products. But when Criterion planned to launch a line of men's toiletries, they were well aware that introducing a line of products unrelated to any product categories with which Criterion was previously associated would present entirely new challenges.

The men's toiletries category is crowded and dominated by established names. The new line was expected to be positioned between the high and low ends of the market. Developing an effective and memorable brand name and visual identity were felt to be essential for the venture's success; this identity needed to operate as the focal point for packaging and advertising strategies. To accomplish these objectives, the company formed a name development committee consisting of the brand's marketing manager, the advertising agency account executive, and the design consultancy's president and called on the other experts, such as research and legal consultants, as needed. Since the design consultancy had extensive experience in consumer product name development, they chaired the committee and supervised the name-generating process.

Challenge

The challenge was to create a distinctive name and visual identity for a new line of men's toiletries, which would serve as the keystone for the initial product launch, penetrate an already saturated market segment, and earn long-term consumer acceptance, a name that could be marketed overseas as well as in the United States.

Analysis and Procedures

Preliminary category and product research conducted by the marketer and the ad agency was analyzed. The products in the line received high ratings among all age and income groups, which suggested that the new brand could earn acceptance beyond the middle market to include a broader demographic profile. This guided the name and visual identity choices.

Identities and packaging of competitive brands in the category were studied in actual retail environments and in TV and print advertising.

The team formulated a clearly articulated strategy and criteria for name development; these criteria served as a framework for generating and evaluating name candidates. It was agreed that the name must communicate excitement and adventure. Some of the other communication criteria were

- quality and product efficacy

- distinctive and appropriate to the category

- easily distinguishable from competition

- easy to pronounce and recall

- good attributes for visual interpretation (design)

- suitable for use overseas and/or translation to other languages

- legally available in the United States and overseas

An examination of names already owned by the company did not uncover suitable candidates. Names were then generated along two directions: descriptive words and terminology (for example, Tasters Choice and Pampers) and fabricated names devised from word roots, prefixes, suffixes, and combinations (for example, Duracell and Kleenex) that project positive imagery and promise of benefits.

In addition, a review of the design consultancy's comprehensive name bank produced a wide range of candidates.

Name generation was achieved by means of three developmental processes:

- a full-day brainstorming session by the team, whose members had discussed the situation with colleagues in their respective organizations and who brought with them ideas and name candidates

- research among a variety of sources that are often fruitful, such as mythology, animals, history, geography, astronomy, and sports

- computer assistance, by the feeding of parts of words and syllables into a computer program that produces a wide range of basic concepts, which then require modification by the consultants to form names that can be pronounced and understood

These processes led to the selection of fifteen candidates from the seven hundred that were generated. Each of the selected candidates was subjected to a preliminary trademark review. Ten name candidates survived.

After more intensive evaluation, the list was pared to four names, which were linguistically checked to ensure that none would have negative connotations in Spanish, French, German, Japanese, and Chinese.

Visual attributes were explored for each candidate. Symbols, typographic treatments, stylized signatures, and other graphic devices and shapes were designed to determine which name offered the best potential for a unique and memorable brand identity. Three finalists were chosen, each adapted to mock-ups of packages—a bottle label, a carton, and a spray can. Each was also evaluated by the ad agency for effectiveness in advertising.

The three candidates, shown in the context of prototype packages and in TV and print ads, were tested among focus groups and in one-on-one interviews with consumers. Respondents including men *and* women were questioned, since the latter are responsible for purchasing or influencing the purchase of more than half of the men's toiletries that are sold.

Based on the results of the interviews, the brand identity candidates and prototype package designs were refined and visually evaluated in actual retail situations amid competition to ascertain shelf power. The project team then ranked the three candidates in terms of how each fulfilled the criteria and selected one name and brand identity concept for recommendation to senior management.

In the presentation to senior management, the team reviewed the criteria and strategy, name selection process, logo and package design exploration, advertising, and research and testing and provided the rationale for the name and design decisions. The team's recommendation received management's enthusiastic endorsement. Legal registration was finalized.

Results

Following the next steps—completion of design for brand identity and packaging, approval of the advertising program, and product and packaging production—the products were rolled out first in two test markets, then nationally. Within one year, the brand captured a 22 percent share of market in the U.S. men's toiletries category and continues to grow as new products are added to the line. After being introduced in the United Kingdom where the new brand is gaining acceptance and is outselling some established products, launches in other countries are on the way.

Insights 5: Avoiding Potential Pitfalls

Background

Competitors, especially in the specialty foods area, have been chipping away at the multimillion dollar franchise that Rockport Baking Company has been building since the turn of the century. Research showed that Rockport's familiar red-and-yellow packages were popular with longtime users but looked "tired" to younger consumers with families who were attracted to other brands whose packaging promised new taste thrills. Since cookies accounted for more than 50 percent of Rockport's sales and an even higher percentage of profits, any change in packaging, no matter how minor, required approval from the board of directors.

The issue of package redesign became increasingly sensitive as the subject moved up the chain of command from brand and marketing managers to senior management. The CEO compared changing the packages of Rockport cookies to redesigning the American flag. But declining sales and loss of market share continued. The investment community no longer recommended the company's stock for its growth potential. Finally, pressured by "the numbers," management realized that they needed to investigate redesigning their dated-looking packaging to stem the decline and regain their market. A search was undertaken for a design consultancy. The marketing and brand managers understood that if the label redesign program failed, their careers at Rockport probably would be over. A package design committee was formed consisting of the senior marketing vice president and two brand managers.

After a preliminary search among several design consulting firms, five of these were invited to present their credentials and relevant experience to the package design committee. From these presentations, the search was narrowed to three consultancies, each with extensive experience in food packaging. Following preliminary orientation meetings with each of the three candidates, proposals were requested and carefully studied by the committee.

In the meantime, the graphic design firm that had created Rockport's award-winning annual reports for almost a decade asked to be considered for the packaging assignment. The design firm's principals had worked closely with Rockport's senior management, who felt comfortable with the relationship. However, their proposal was rejected by the package design committee, who fully appreciated the firm's excellent reputation and design capability but felt that the firm's lack of experience in packaging was evident in the proposal's failure to address several critical issues essential to the package design investigation. The firm continued to intensely lobby senior management, and the CEO of Rockport overruled the committee's selection, instructing it to move ahead with the annual report firm.

Challenge

The challenge was to explore and make recommendations for updating packaging, enhancing taste appeal and merchandising effectiveness of the Rockport cookie lines to attract younger consumers while considering established recognition and image equities to retain current users.

Analysis and Procedures

Senior management's intervention in the choice of the design consultant exacerbated an already sensitive issue. The committee made every effort to thoroughly brief the design firm and to review findings that were culled from several research studies and retail audits. Since the design firm brought little experience in package design to the situation, the committee made extraordinary efforts and spent considerable time to "tutor" the design firm with regard to packaging issues.

Knowing senior management's exceedingly conservative attitudes about packaging changes, the design firm explored a range of design treatments, all of which were very close-in to the existing packages. The major change was enlarging the Rockport logo and making the typography for product descriptions more legible. The logo styling itself and the overall layout, including the predominantly red-and-yellow color scheme, were left unchanged on all the design firm's concepts.

Before presenting the results of the design investigation to the package design committee, the head of the design firm showed the design explorations to the CEO, who was pleased that the existing format had not been violated and modified only minimally. Then the work was presented to the package design committee.

Word was passed down that senior management already had seen and liked the designs. The committee was stunned. While maintaining a relationship to the traditional Rockwell look, they had hoped for recommendations that would substantially strengthen appetite appeal and communicate excitement to a younger target audience, none of which was evident in the design firm's explorations. The committee felt that the range of concepts recommended by the design firm was much too narrow. Nevertheless, in view of senior management's reaction, the committee decided to go forward with consumer research as initially planned. Two of the package design concepts were mocked up, applied to the most popular cookie varieties in the line. An independent and highly respected research organization was called in to ascertain how the revised designs were perceived by consumers compared to the existing packages.

The findings confirmed the committee's concerns. The revised label treatments tested poorly. Respondents saw little difference between the existing packages and the new designs and expressed opinions that the packages looked as if the cookies would not taste as good as the products in the existing packages. When tested against competition, the revised Rockport designs were described as "traditional" and "reliable" but were not seen to communicate great taste and adventure to younger consumers.

Results

This classic case of choosing the wrong consultancy for the wrong reasons should have been avoided. The CEO was now in the embarrassing position of having to reverse his own decisions and to approve funds for additional package design explorations by yet another package design consultant. Responsibility for this unhappy and unproductive experience must be shared by *both* marketing management and the CEO. The CEO and his associates at the top, after giving the package design committee the responsibility of researching and recommending a design consultant, should have displayed more confidence in the marketing group by allowing them to conduct the redesign program according to proven package development practices. On the other hand, marketing management made the mistake of not recommending, even insisting, that senior management meet the three candidates they had selected and listen to their proposals.

Insight 6: Creating Brand Identity and Packaging for Global Marketing

Background

Kenton/Palmer, a leading marketer of several oral hygiene products, planned to introduce its bestselling K/P+ line in selected overseas markets. In the United States, marketers of lower-quality toothpastes, gels, toothbrushes, and related items imitated K/P+'s packaging, which confused consumers and cut into sales. To resolve this problem in the United States and to preclude similar conflicts overseas, management decided that K/P+ needed more proprietary brand identity and packaging graphics that could be legally protected and give all K/P+ packages a unified look in cross-border marketing regardless of name and language.

Challenge

The challenge was to design a brand identity and packaging format that speaks with one voice in all markets, is legally defensible, has built-in flexibility to accommodate package design equities at local levels, and is adaptable to retailing, cultural, and linguistic differences in various global locations.

Analysis and Procedures

Moving an established brand from U.S. to foreign markets is a sensitive process fraught with potential pitfalls. Expertise in design and

packaging is not sufficient; success depends largely on understanding of retail situations in various countries around the world,
as well as cultural and linguistic nuances that often play critical
roles in global package development. These may include usage customs, words, colors, typographic styles, package shapes, sizes, and
structures. What works for a brand in the United States may not
be equally acceptable in France, Brazil, or Korea.

To help explore options, management retained a design consultancy with experience in multicultural communications. The
designers reviewed packaging and brand graphics of comparable
products in the targeted overseas markets as well as in the United
States to better understand how these competitors are visually
identified and how they function in various regional markets
around the globe.

K/P+'s category and product research was studied and orientation meetings held with marketing management in the United
States and marketing directors from several key countries in
which K/P+ would be launched. The criteria for positioning the
brand were formulated from a global perspective with special
attention to ensuring cross-border recognition and ensuring that
identity and packaging modifications could be easily implemented
transnationally.

To minimize cloning of the K/P+ graphics by competitive
global brands, the designers focused their creative efforts on achieving a distinctive brand mark that would resist imitation attempts
because of its proprietary character and that would not project any
negative connotations in global markets.

The problem was solved by restyling the K/P+ brand logo,
placing it into a unique shape, and developing a distinctive package color. In this way, the entire package became a recognizable unit
that satisfied the issue of global trade dress ownership. Although
the product descriptions would be translated into languages other
than English outside the United States and the United Kingdom,
the brand identity and package colors stayed the same, thereby supporting the brand's recognition everywhere.

Once approved by the key countries, the new identity and package design program received enthusiastic support from marketing
managements everywhere.

Results

By devising a skillful solution to a complex global marketing and legal challenge, Kenton/Palmer was able to launch K/P+ in Europe, Latin America, and Asia, where it has captured key positions within less than two years. Kenton/Palmer is now a major factor in the worldwide sales of oral hygiene products and plans to expand its global presence to other brands in its product arsenal.

Insight 7: Launching a New Brand

Background

Voyager Chemical Corporation, an international industrial chemical manufacturer, decided to establish a consumer products division that would leverage the company's experience in basic chemicals and open up new business opportunities. Voyager's management felt that a number of their products could be modified to be sold over the counter at retail. After investigating a range of opportunities, a category was chosen with which Voyager could feel comfortable: lawn and garden products.

To acquire retail marketing know-how, Voyager purchased a small regional company that sold seeds and lawn care products. A team with representatives from this company and Voyager was organized to operate the new venture.

Challenge

The challenge was to develop lines of grass seed and lawn care products for distribution throughout the United States that would generate attention from trade customers and consumers through the use of distinctive user-oriented packaging. Although some advertising support would be provided, the brand would rely primarily on packaging and merchandising in garden and hardware stores, mass-merchandise retailers, and supermarkets.

Analysis and Procedures

Creating a totally new brand without concern for retaining established equities in name and graphics always offers exciting opportunities. Marketer and consultant can act in more imaginative ways and explore risks that would not be prudent when updating or repositioning an existing brand.

To explore exciting new directions geared to the retail situations of lawn and garden products—mostly garden centers and hardware outlets—the marketing team retained a full-service design consultancy with experience in supporting new product introductions. A six-month timetable was set so the product lines could be introduced at a major industry trade show.

Criteria were defined for the brand name and product attributes. In this case, the team resolved the name issue quickly by selecting one that the company already owned and that met the criteria, MagiCare. This name also would be suitable for diversification into other home care categories that Voyager planned to pursue in the future.

Logotype design was completed within four weeks. The Magi-Care name was styled to read as one word that combined the spelling of two words, *magic* and *care*, with a distinctive letter C that served as a ligature and visual focal point.

Management decided to market the MagiCare products under a single brand umbrella but divided the line into two distinct subcategories: products that *nourish* lawns and gardens and products that *prevent* lawn and garden problems. The consultants recommended that rather than using subbrand names, the products be identified generically, such as Crabgrass Control, Shady Growth Seeds, and so forth, in order not to detract from the MagiCare brand name.

To position MagiCare as a superior brand and distinguish the products from competition, dramatic-looking packaging was created. Full-color photographs of lawns and gardens covered the entire front panels of bags and containers and to demonstrate the beauty and luxuriant appearance resulting from treatment with MagiCare products. Since extensive consumer research showed that most consumers lacked in-depth knowledge of how and when to treat lawns and gardens under various climatic conditions, copy on side and back panels was carefully developed by lawn and garden care specialists to educate the consumer.

To further enhance the shopping experience in garden centers, the designers developed special MagiCare racks that were easy to assemble at the point of sale, displayed the MagiCare products in an attractive manner, and made it easy for the consumer to find the desired products. The designers also developed consumer literature that was visually coordinated with the packaging and advertising to pick up the look and theme of the brand's design and packaging features.

Results

When the MagiCare brand was introduced at the industry trade show, the packages won "Best of Show" for new products and packaging and thus brought immediate attention to the new brand, giving it a tremendous sell-in advantage. Demand for the product line at garden centers was instant and exceeded all expectations by the management of Voyager Chemical Coporation. While the packaging has caused competitors to examine their own practices, Voyager is introducing new products in the category to strengthen its image as an innovative marketer of consumer lawn and garden chemicals.

Insight 8: Merging Two Established Brands

Background

Crown Consumer Products, a manufacturer and marketer of a wide range of quality plastic products, was particularly well known for a brand of high-quality bathroom accessories, including soap dishes, toothbrush holders, tissue boxes, mirrors, and other related items. The company also marketed another, more trendy-looking line under a separate brand name. This brand was marketed at a lower price-point, even though the products were of similar quality to their top brand. Despite the availability of a lower-priced brand, the premium brand was outselling the secondary brand line by a considerable margin. This created a situation where some retailers resisted carrying both lines, favoring the higher-priced brand, which was more popular and resulted in higher profit for the retailer. They downplayed the secondary brand to a point where Crown was losing money maintaining the brand. As Crown did not want to give up the secondary brand, management decided to merge it with the premium brand.

When the Crown sales force heard about the plan, they did not like it, nor did those retailers that were carrying the lower-priced brand. Neither did they want price increases that they antic-ipated as a result of the merging of the two brands. Crown management decided that one way of solving the problem was to develop packaging for the accessories that would reflect the equity of the premium brand, concurrently promoting the benefits of the lower-priced items, and offer these at a compromise price-point.

After interviewing several design consulting firms that had proven experience in marketing and design, which was critical to dealing with this situation effectively, one firm was selected and asked to participate, together with the marketer's ad agency, in the strategic planning meetings.

Challenge

The challenge was to create a packaging system that would achieve a single, proprietary look for the brand, building on the positive reputation of the premium brand and merging it with the trendier secondary brand. To appease retailers as well as the internal sales staff, this needed to be accomplished quickly and take into con-sideration some of the equities of both brands.

Analysis and Procedure

One-on-one interviews by the consultant with some of the com-pany's staff (especially the sales staff) and telephone interviews with purchasing directors of some of Crown's key accounts set the stage by making those who questioned the brand merger feel that they were involved in the decision-making process. This reduced some of the psychological pressure on the project and set the stage for more relaxed and deliberate decisions. The consultant suggested four alternatives:

- Use the brand name and logo of the primary brand and drop the secondary brand. This "cold turkey" approach would probably cause strong reaction but could be handled by simply riding out the storm.

- Use the brand name and logo of the primary brand, but create packages in the trendy style of the secondary brand.

- Use the brand name and logo of the secondary brand for part of the line but endorse it with the premium brand name.

- Generate an entirely new brand name and logo, thus neutralizing any leanings for or against one of the brands.

After several days of deliberation, brand managers and the consultant mutually agreed to recommend to Crown's top management a combination of two of the options: using the premium brand name for the bestselling products, using the secondary name with the endorsement of the premium brand for selected, less popular items, and styling the packages in a trendier manner.

Results

The packages resulting from this approach satisfied all parties concerned with merging the two brands. By using the premium brand name as an endorsement, the equity that this popular brand had created over the years was leveraged. Using the name of the secondary line for part of the line prevented this line from sudden death and pacified the retailers who had carried it. Introducing a more contemporary, though still high-quality, look got the products additional attention from the young generation of new homemakers and gave the entire brand more breadth. Pricing some products slightly below the premium brand ensured a good price/value perception from retailers and their customers.

Thus the merger of the two brands created new opportunities for Crown Consumer Products and resulted in achieving even greater emphasis on the brand's premium image at retail than it had initially.

Insight 9: Creating a Second-Tier Upscale Store Brand

Background

Shining Marketing Corporation, the holding company for a chain of more than six hundred mass-merchandise outlets located primarily throughout the southeastern United States, planned to expand their product mix by adding an *upscale* store brand to their already-existing Great Buy private label. Known throughout their sphere of operation for their superior service, Shining was able to attract a growing number of higher-income consumers and felt that the time was right to offer a second-tier, premium store brand covering selected food, household products, health and beauty aids, hardware, and electronic accessories.

There are, of course, many supermarket and mass merchandisers who offer store brands. Several of these have second-tier, upscale lines that offer superior quality at slightly higher prices. To differentiate themselves from these, Shining wanted to make available products that were not just better quality but that were *different* from what other store brands offered, including their own Great Buy brand, and that would *not be available* at any other stores. To facilitate this ambitious strategy, Shining initiated the program by acquiring the services of several consulting and procuring specialists, including

- a consultant specializing in product development
- a firm for sourcing, procuring, and processing products and packaging
- a brand identity and package design consultant for naming and package design development

Challenge

The challenge was to create a brand name and a store brand packaging program that didn't look like any other for products that are different and better than any others and to position the new brand to compete head-to-head not only with the store brands of other mass merchandisers and supermarkets but also with major nationally advertised brands.

Analysis and Procedures

To set into motion and expedite the development of the new brand and all that was required to accomplish the marketing objectives, Shining arranged to have biweekly strategy meetings to discuss the status of the program; review various developments that occurred during the previous weeks; discuss the progress of developing, securing, and processing selected products; and plan for new product and package development during the following weeks.

The first step to move the program forward was deciding on the type of products to be included, the price/value relationship of each vis-à-vis other products in their own stores, store brands of competitive stores, and national brands. Next was to develop a name for the brand and packaging to communicate the brand's positioning strategy to the consumer: premium-quality products available only at Shining stores and at lower prices than national brands.

The creation of the name for the new brand was as unique as the program itself. While the objectives for most brand names are to be as different from other brand names as possible, the criteria set by Shining was that the name for their second-tier upscale brand should sound familiar and lend itself to being used as an integral part of the product communication.

After reviewing a large number of name candidates and checking their availability, the name that surfaced as everybody's favorite was Superior. This name was thought to be most appropriate because in addition to being short and easy to remember, it sounded familiar and supported an image of high quality. Most important, it could function as part of the brand's product communication, for example, Superior orange juice, Superior videotape, Superior light bulbs, and so forth.

Having approved the name, the design consultants developed an eye-catching logo that further supported the upscale image of the brand. The logo was geared not only to stand out on packages in shelf display but to be used on shelf talkers, internal signage, in-store displays, and other promotional needs, thus calling attention to and promoting the brand throughout the stores.

Meanwhile, the biweekly strategy meetings went into full swing, attended by all individuals involved in launching the new brand, including representatives for product development, product sourcing, manufacturing, marketing, and packaging and frequently even including the CEO of the company, emphasizing the urgency of the program.

Thus, as products were brainstormed and produced at record speeds, subject to Shining's strict quality control, package design for each product came under special pressure. This was facilitated, however, by the fact that design approvals were the responsibility of the small group of project participants, most of whom were available at the strategy meetings, which simplified and expedited design approvals tremendously.

To further promote the introduction of each new Superior product, banners, balloons, and in-store posters boldly and cheerfully drew attention to each launch, and numerous end-aisle displays stocked with new Superior products greeted the shoppers as they entered the stores.

Results

The coordination by everyone connected with the introduction of the new upscale brand and the multiple launches of various products that followed acted as a powerful stimulus that catapulted the brand to the forefront of the stores' activities. The development of attractive packages was expedited by abbreviating the usual requirement of multiple layers of corporate approvals (which has the effect of watering down creativity in exchange for "security"). The packaging not only succeeded in supporting the store's strategy of communicating premium quality at a lower price but won numerous accolades from shoppers and several prestigious awards from industry and professional sources.

Insight 10: Balancing the Demands of Two Competitive Distribution Channels

Background

Lumiere Corporation is a leading U.S. manufacturer of table lamps, from inexpensive metal-base structures to high-priced collector's items featuring hand-decorated porcelain bases. The line consists of five brands, each representing either a style such as traditional, modern, or decorative, or a price-related tier. The lamps were traditionally available at better department stores, but more recently, mass-merchandising outlets expressed the desire to carry them as well and threatened to turn to overseas sources if Lumiere were not willing to supply them with their products. After extensive deliberations by Lumiere's management, it was decided that rather than fighting what promised to be a losing battle with the powerful mass merchandisers, Lumiere would make their entire line of several brands available to both department stores and mass merchandisers. However, in order not to jeopardize their long-standing relationships with department stores, the products would be packaged differently for the two competing distribution channels.

In the past, Lumiere admitted to itself, management had paid little attention to how their products were packaged. The lamps sold well, and packaging was not on Lumiere's priority problems list. They were shipped in corrugated containers that, though somewhat decorated, functioned concurrently as shipping containers and store displays. In fact, until now, packaging had been the responsibility of the purchasing department, who relied on their suppliers to provide packaging graphics along with the containers. No one at Lumiere had ever used a package design consultant, though their ad agency had tried to persuade Lumiere on many occasions to consider doing so.

Now, under pressure of restructuring the architecture of their entire line to satisfy the two competing marketing entities, Lumiere's marketing v.p. was assigned to seek the assistance of a reputable, marketing-knowledgeable package design consultant. After a brief search, one of the leading consultancies was asked to analyze the situation and make recommendations as to how to handle the packaging dilemma.

Challenge

The challenge was to develop a dual brand architecture that would ensure consumer awareness of the core brand (Lumiere) no matter where displayed and clearly differentiate between product lines sold in department stores and those available at mass merchandisers. Beyond these criteria, the overall objective of the packaging program was to capitalize on the positive attributes of the Lumiere imagery while identifying and enhancing the individual brands and the products within these lines.

Analysis and Procedure

As a first step, the design consultant felt that since very little meaningful information about consumer attitudes towards these products was available, it would be necessary to do at least some informal one-on-one interviews with consumers as well as store personnel. This was done in the context of retail audits in several major department stores and mass-merchandising outlets. These activities brought to light that consumers clearly differentiated between the two primary channels of distribution: mass merchandisers and "class" merchandisers (department stores). Packages serve marketing in two ways in these surroundings, as "silent presenters," when cartons are stacked to form a mass display, and as "silent salesmen," when packages provide brand appeal and product information. Since all of Lumiere's packages were corrugated containers, these would not satisfy the need for differentiating between mass-merchandising and department store merchandise.

In addition, through research it became clear that although the Lumiere line had been marketed under five different brand names for a long time, few of the consumers interviewed were aware of this and those that were did not understand the reason. As each brand faced different communication issues with regard to styles and price, the consultants felt it was important to develop packaging that would solve issues that were central to Lumiere's new marketing strategy. The new packaging must provide for

- strong overall corporate identity to bring greater consumer awareness to the Lumiere corporate name, creating an image of quality and trust for all Lumiere products

- more meaningful differentiation between each of five subbrands

- clear differentiation between products sold at mass-merchandising outlets and those available in department stores

- attractive carton graphics on the department store packages to capitalize on the attractiveness of the lamps, enhance in-store displays of Lumiere lamps, and make the packages more suitable for gift-giving

- simpler but improved graphics for the mass-merchandising packages to differentiate them from department store merchandise

- copy to help the consumer select the right lamp and to point out special features and attributes as compared to cheap imitations

Based on these criteria, which were approved by Lumiere's management, the designers developed a packaging system for three of the five brands to be completed in time for the Christmas holiday season. The resulting packaging scheme featured and emphasized a constant Lumiere logo on all packages. Full color photographs covered at least one entire main panel of each department store package showing the lamps in attractive home settings. Black-and-white photographs on soft-pastel-colored backgrounds appeared on the mass merchandising packages (the colors doubling as color coding to differentiate the brands). Even the substrates were improved on the packages for both channels in order to achieve the desired print quality.

Results

With the new packaging systems in place well before the critical Christmas season, the introduction of the new graphics was well received by both channels. The mass merchandisers, having Lumiere lamps available in their stores for the first time, did a booming business that exceeded even their high expectations. The department stores, seeing in the full-color cartons new opportunities for effective displays, also did very well, and the attractive packages allayed their previous objections to opening the Lumiere market to mass merchandisers. Taking into account some minor modifications that resulted from the sell-in of the three initial brands, the consultants now proceeded to adapt the system to the remaining two brands and to create guidelines for future packaging of additional brands and products.

Glossary

This glossary is meant to clarify words and expressions that are common language among marketing and design professionals but with which some of our readers may not be familiar. Although some of these terms may have different meanings depending on usage in a variety of other disciplines and applications, this glossary is meant primarily to define the terms for their relevance to brand identity and package design.

acrylic a plastic material used for producing solid bottle models

aerosol container a container that employs a propelling agent to discharge its contents through a valve

animatic a preliminary concept presentation of a TV commercial created for client approval prior to full production

audience consumer or trade groups to whom the product and package is addressed

bleed the print area beyond the cut edge or fold of a package, required when an image or color covers the display panel from edge to edge

blister pack a package consisting of a plastic, transparent form secured to a cardboard or plastic surface, containing and displaying one or more products

bottleneck the narrow upper portion of a bottle

boutique a separate section within a (usually large) retail store selling specialty products, such as bakery goods, gourmet cheeses, or ice cream

brainstorming spontaneous conceptualization by a group of participants for the purpose of generating a wide range of ideas

brand a name that distinguishes one product or product group from another

brand elasticity the extent to which a brand name can be expanded to a variety of products or product categories without diluting the brand value

brand equity the values that a brand name has established among consumers in the marketplace

brand extension the addition of products under the same brand name

brand frame the shopping environment of a brand and the manner in which the brand functions within in it

brand identity the total of all elements of a brand, from name to visual and verbal communications, that differentiate one brand entity from all others

brand image the impression that is formed about a brand and its products in the consumer's mind

brand personality the sum of qualities and impressions communicated by the brand

brand recognition the ability of a brand to trigger awareness and be remembered by the consumer

brandmark a word or symbol or a combination of both that distinguish one brand from another

burst a visual element on a package, usually temporary, that draws attention to a special event or development, such as a new or improved product, a unique product feature, or a special sale

buyout the acquisition and ownership of all rights to a photograph or illustration

captive designer a term used to describe a designer employed by a supplier to create packaging structures and/or graphics exclusively for products produced and marketed by the supplier

carrying charge the financing cost for purchases made by the designer on behalf of a client, often added as a percentage to the actual purchase costs

category the classification of products that have similar functions, meanings, or objectives

category analysis an examination of various factors within a product category that influence sales positively or negatively

caution copy the words that call attention to proper handling and/or usage of a product

checkerboarding a term describing the interspersing of similar-looking private labels with the packages of leading brands, creating the display effect of a checkerboard

chemical migration the usually undesirable movement of chemical substances from the container into the container's contents

child-resistant packaging packages with features that inhibit a child's ability to open a package and access its contents

clam pack a one-piece plastic structure that folds and closes in a manner resembling a clamshell, usually transparent to display the entire contents

cloning a term for imitating another package design, most often that of a leading brand

close-in (redesign) an evolutionary package redesign that retains a close visual and/or structural resemblance to an existing package for reasons of image continuity

color coding the identification of product varieties by means of colors

color separation the separation of the colors of photographs and artwork through optical or electronic means to produce printable color components

color trap a slight overlap of adjoining colors during the printing process, created to avoid undesirable spaces between the colors

confidentiality agreement a legal agreement by the designer to not divulge or discuss information to which the designer has access with anyone not directly connected to an assignment

consumer feedback research to determine consumer reaction to new design concepts and/or existing packages

contoured can/bottle a can or bottle that is shaped or sculptured rather than straight sided

copy the verbal elements appearing on a package

copycat (design) a term for a package that looks like that of another brand

copyright the legal protection of rights to ownership of a design or a specific design element

core brand the primary name of a brand, essential for maintaining brand recognition and image transfer for line extensions or flanker brands

corporate identity the total of all elements of a corporation that work together, from name to visual and verbal communications, and that differentiate one corporate entity from all others

corrugated board a packaging material consisting of a fluted (corrugated) paper core usually sandwiched between two paper liners to form a strong entity used for product protection and shipping containers

corrugated container a container made of corrugated board

cylinder a round form used to hold rotating printing plates

design brief a document, usually prepared by the marketer, outlining background, criteria, and all other factors pertinent to a design assignment

design control manual a document to help assure the consistent and correct implementation of all elements of a design program, including brand identification, verbal and visual elements, copy, colors, and if applicable, structural components

design criteria the standards, rules, parameters, and objectives for a design program to guide and ultimately measure the design development

design exploration the search for design directions that will fulfill given design criteria

design finalization a general term describing any type of final preparation for production, such as working drawings or blueprints (for structural design) and/or mechanical artwork and photography (for graphics)

design guidelines a document for implementing a structural or graphics design program in an approved and consistent manner

design modifications/design refinements changes made to designs or design concepts based on consumer research feedback, market conditions, or judgmental decisions

die a metal tool used for cutting paperboard, shaping metal cans, and forming extruded plastics

die imprint an imprint made by a metal die on a substrate to indicate the dimensions, contours, scores, and print areas of a package for mechanical art preparation

drawn can a can formed by pushing a flat sheet of metal through a circular die

drop strength a measurement to determine resistance to damage of a package dropped from predetermined heights, used especially for testing shipping containers

dry offset printing a printing method most often used for transferring graphics to aluminum cans and plastic tubs

dump display a random display of packages in a basket or display unit, often used for special sales and to encourage consumer purchase

electronic art artwork prepared by electronic means, that is, on computers

electronic design control manual a digitized method of design implementation that determines placement, styles, sizes, and colors of package design elements electronically (also see *design control manual*)

end cap a display of products at the end of a store aisle to promote or draw special attention to a product or product line

endorsement support by a company for a brand or support by a brand for its line of products

equity the value that a brand, a package, a package element, or a color has established in the marketplace

Eye-Tracking a research methodology based on an optical scanner that tracks the respondent's eye movement across a package or a package display

film a thin, flexible plastic material

finish the end of a glass or plastic container to which a closure is applied

flanker brand a secondary brand, related to the primary brand, marketed under its own brand name

flexible bag/pouch a package made of flexible material such as film, paper, or foil

flexography a printing method using raised images to print on corrugated board, film, and paper, sometimes more economical than other printing methods as a result of using less expensive rubber or polymer printing plates

flip cap a package closure that can be flipped open to access the contents

focus group interviews a consumer research method for probing consumer attitudes using an open exchange of ideas among small groups of consumers, guided and controlled by a moderator

foil a thin metal sheet used to protect products such as food against spoilage or as a decorative element laminated to the surface of paperboard or plastics

folding carton a paperboard container delivered flat or folded and erected on the product manufacturer's packaging line to form a carton

food stylist an individual specializing in preparing food for photography

freelancer a person providing services on a temporary basis

gable-top carton a paperboard container folded at the top to form a gable-shaped closure that can be opened for pouring, most often associated with dairy products and juices

graphic design the composition of visuals that decorate the surface of a package and provide information about its contents

graphic format the arrangement of the visuals on the surface of a package

gravure a printing method using a metal cylinder into which cells are etched or engraved to hold various amounts of ink, most often used for long runs of packages because of the durability of the cylinders and high speed of printing

grid a meshlike background on which graphic elements are arranged to indicate precise measurements

Grüne Punkt (Green Dot) the Packaging Ordinance on Waste Avoidance, requiring German manufacturers and distributors to take back discarded plastic packages for recycling

halftone art the conversion of artwork into tiny dots of varying sizes that, when printed, transmit to the human eye the impression of dimensional images

harmonize to create visually corresponding though not necessarily identical packages, as may occur with global, multi-language packaging programs

heat transfer printing a decorating method that uses heat to transfer an image from a preprinted film onto the package substrate, frequently used for decorating plastic bottles

hot fill a technique for filling bottles or jars with liquids or food in a heated state for the purpose of product sterilization or preservation, creating special structural requirements for the hot-filled packages

hot stamping a decorating method that delivers an image via a transfer film onto the package substrate, resulting in a thick, glossy image

icon a symbol that identifies a "character" for the purpose of recognition or association with a familiar object or personality

image the impression that is formed about company, brand, or product in the consumer's mind

image communication the visual and verbal transferral of an image to the consumer's mind causing the consumer to form an impression about the company, brand, or product

indemnification clause a legal term for specifying responsibility for the costs involved in legal actions relating to an assignment

intellectual property the legal ownership of inventions or proprietary business property, including patents, trademarks, trade dress, and copyrights

knockoff a term for copying or imitating the package design of another (usually leading) brand

label a sheet of paper or plastic containing brand and product information that is affixed to a package

lamination a combination of several layers of materials joined together to form a material not available in a single layer for the purpose of achieving superior strength, product protection, or printing surface

layout the composition, arrangement, or concept rendition of a graphic design

letter of agreement an informal, usually brief understanding having legal validity, outlining the mutual obligations and responsibilities of the participating parties during and after the execution of a design assignment

letterpress a printing method using metal or polymer printing plates with raised images to transfer viscous ink onto the package substrate

leverage to utilize the positive imagery of a brand or product to expand or broaden a brand or a line of products

line extension the addition of products or product varieties to an existing brand line

logo/logotype the graphic identification of a company or brand, usually a signature or symbol or a combination of visual devices

machinability the capacity of a package to be set up and filled using available machinery or machinery designed specifically for a package

mass merchandiser a large retailer, usually providing a wide variety of products at low price-points

master brand the primary name of a brand, essential for maintaining brand recognition and image transfer for line extensions or flanker brands

matched color a color matched to specifications (as opposed to a color derived from the combination of process colors)

mechanical artwork the final artwork for print production

mock-up a scale model of a package, often including the surface graphics, developed for the purpose of visualization and design approval

model (1) a packaging concept created three-dimensionally, using wood, acrylic, paperboard, or other materials; (2) a person who poses for a photographer or illustrator

model agent a person representing professional models in terms of business presentations and contractual negotiations

mom-and-pop store a term used to describe a small, privately owned store

neckband a label applied to the neck portion of a bottle

no-name brand inexpensive merchandise, usually of lesser quality, in packages that feature only *generic* product identification

noncompete clause a contractual stipulation that the designer cannot work for a competing company or on a competing brand for an agreed-upon period of time during and after an assignment

Nutrition Facts panel the mandatory nutritional label or panel required on all packaged foods in the United States by the Nutritional and Education Act of 1990

nutritional information the identification of the contents of food and beverages in terms of calories, protein, fat, cholesterol, and so forth per serving

odor transfer the usually undesirable transmission of odors from the package material to the product inside or vice versa

offset printing a printing process using the immiscibility of oil and water to transfer images from a metal plate to a rubber blanket that offsets the image onto the label or package surface

orientation meetings sessions at the beginning of a design assignment with the client's marketing and technical personnel, the ad agency, and sometimes suppliers, to provide in-depth marketing information, technical input, and strategic objectives to the design consultant

OTC (over-the-counter) a term describing pharmaceutical products available at retail without a doctor's prescription

out-of-pocket expenses expenses incurred by a design consultant for purchases such as competitive products, photography, transportation, and hotel accommodations in connection with a design development program

own label a term describing a private label program, used mainly in Europe

package any container intended to hold, protect, transport, store, and identify a product

package design development the entire process by which a package is created and prepared for production

package form the type or shape of a package

package graphics the visuals that decorate the surface of a package and provide information about its contents

package structure the shape, material, and functional components of a package

package supplier a manufacturer or purveyor of packaging and/or packaging materials

packaging line the machinery at the product manufacturer's plant on which packages are erected, sorted, filled, closed, labeled, palletized, or otherwise processed

pallet a platform, made most often of wood or plastic, for combining and stacking a number of packages, creating a level unit to facilitate distribution, transportation, and storage

palletizing the process of combining and stacking packages on pallets

paperboard paper material of various thicknesses that can be shaped and assembled to form a package

paste-up an art preparation technique by which typographic and other visual material are pasted up on a cardboard in preparation for print production

Peg-Boarding a display method by which products are hung on Peg-Boards

PET (polyethylene therephthalate) a popular plastic material used for various types of plastic containers, primarily beverages

phase a step, usually describing a particular development segment of a design program, such as concept explorations, design refinements, research, or finalization

plastic a synthetic material capable of being formed, shaped, and molded

post-design research research to probe consumer reaction to a variety of brand identity and package design concepts that guide refinements and/or assist in design selection

pre-design research research to determine consumer attitudes toward a product, a product category, or specific packaging issues, helpful for guiding design explorations

pre-production the process of color separation for printing and other types of preparation for package production

price-point the price of a product or the price level of a product category

price/value the relationship of the cost of a product to its perceived worth

primary package the package that contains the actual product

printout electronically prepared final artwork used for proofreading and client approvals

private label a line of products marketed by a retailer under the retailer's own name (also referred to as *own brand, store brand, house brand,* and *exclusive brand*)

process colors the basic colors used for printing full-color images: cyan (a particular blue), magenta, yellow, and black (or a dark color close to black)

process tint a lighter shade or gradation of a color

product description the words and visuals that define the
    product's characteristics and benefits

product line a series of related products marketed under a single
    brand name

product personality the distinctive, individual identity and
    quality of a product as perceived by the consumer

production art the final artwork for print production

production art file electronically prepared final artwork saved on
    a hard drive or disk for future use and reference

proof a trial reproduction of final artwork produced either on
    specifically prepared material or on the actual printing press
    for checking the correctness of the copy and visual images

proposal a document prepared by the design consultant that sets
    forth an outline of a design program, describes each step
    (phase), and specifies fees, incremental costs, and timing

reading stat a printout of typeset copy to check its correctness

recognition requirement the need for recognizing and
    identifying a brand or product by the shape and/or the
    visuals on a package

reflective art mechanical artwork prepared in pasteup form as
    opposed to electronically prepared artwork (also see *paste-up*)

residual the provision for additional fees for photography if the
    same photograph is used for other than the originally
    contracted application

retail analysis an in-depth study of the conditions existing in a
    specific product category or in the retail environment in
    general

retail store any store offering consumer products

retainer a contract between a marketer and a designer by which
    the designer provides an agreed-upon volume of work for a
    specified period of time and fee arrangement

retouching a method of altering original photographs, such as adding or deleting visual elements, removing blemishes, or modifying colors

rotogravure (see *gravure*)

sans-serif without a serif (see *serif*) used to describe a typeface usually projecting a modern or bold "personality"

screen printing a printing method using a mesh that regulates the amount of ink transferred to the printing surface

screw cap a plastic or metal closure that is screwed onto the finish of a plastic or glass container

secondary package an outer package containing a primary package, added for purposes such as product protection, transport, imagery, display, or sales promotion

sell copy copy that promotes the assets of the product in the package

separator a person or company preparing artwork for print production, including color separating photographs and color artwork (see *color separation*)

serif a horizontal stroke added perpendicularly to the top and/or bottom of a typographic character to give it a unique aesthetic "personality" (also see *sans-serif*)

setup carton a carton that is delivered already formed, as opposed to a folding carton that requires setting up (see *folding carton*)

shelf life the time span during which a packaged product remains in satisfactory condition, especially applicable to food, beverages, and pharmaceuticals

shelf talker a label, flag, or poster attached to a store shelf to promote a product, special sale, or price advantage

shrink film a film that is heated to shrink tightly around an object to hold it in place and/or protect the contents

signature the graphic treatment of a company or brand identification, a term interchangeable with *logo* or *logotype*

single-cavity mold a preliminary mold for a plastic or glass container for the purpose of checking volume and appearance

SKU (stock keeping unit) a term used by manufacturers and retailers to identify individual packaging units for inventory purposes

specialty store a retail outlet specializing in a particular type of merchandise

speculative design design services supplied by a designer to a marketer at minimal or no fees in the hope of gaining an assignment

stock photography nonexclusive photographs available to any user, usually for a lesser fee than custom photography

store brand products marketed by a retailer under the retailer's own name (also referred to as *own brand, house brand, exclusive brand,* and *private label*)

storyboard a series of concept renderings of frames representing the sequence for a television commercial

structural design the design of shape, materials, closure, and other physical package components

subbrand a secondary brand, usually endorsed by a core brand

substrate the packaging material and/or material surface that forms the basis onto which images are printed

symbol a graphic element associated with a company or a brand, often part of the logotype

tachistoscope (T-scope) a tool used for consumer research that measures the response time to packages or package displays exposed at very brief intervals

tamper resistance the property of some packages and closures to resist access to products by individuals who may intend to make them harmful to users

target audiences consumer groups whose attention to a brand or product is desired by a marketer

text the wording on a package

tier a term used by marketers and retailers describing levels of product quality or pricing

topload strength the limits of resistance to collapse of a package when stacked during storage or transportation

trade dress the overall look of a package consisting of size, shape, material, and all the visual and verbal elements that contribute to its identity

trademark a word or symbol or a combination of both that distinguish one company or brand from all others

transparency a transparent film containing a photograph or image that can be scanned and color-separated to obtain a printed image

typeface a form and style of the letters of the alphabet

typography the arrangement of type used for printing and other kinds of communication

U.P.C. (universal product code) a ten-digit bar code on packages that identifies a product when electronically scanned at the retailer's checkout counter, registering the posted price for the product and, concurrently, controlling inventory management

viscosity the resistance to the flow of liquids

volume the bulk or amount of product in a container

warning copy wording identifying a potential problem or danger regarding the contents of a package

working drawing the finalized rendering of a package structure, detailing all contours and material dimensions and specifying design details to guide the package manufacturer in the development of production blueprints

working model a preliminary, handmade, three-dimensional model of a package that actually functions as though it were a manufactured container

work-for-hire a legal term describing work that is done by a designer who is obligated to surrender all concepts developed in the context of an assignment to the marketer as if in the employ of the marketer

Bibliography

Managing Brand Equity	David A. Aaker, The Free Press, an Imprint of Simon & Schuster, New York, 1991
Building Strong Brands	David A. Aaker, The Free Press, an Imprint of Simon & Schuster, New York, 1995
"Consumer Communications: Shaping Positive Perceptions"	Dr. Wolfgang Ambrecht, Head of Marketing Communications of Products and Services, BMW, Presentation, The Corporate Image Workshop, The Conference Board, New York, 1995
"Trends in Retail: The Growth of Merchandising Services"	Donald L. Berry, SPAR Marketing, Inc., *Public Relations Bulletin*, Bloomfield, MN, 1995
"Brand Extensions: Lessons of Success and Failure"	Nadia Bunton, *Brandweek*, June 28, 1993

Private Label Marketing in the 1990s   Philip Fitzell, Editor, Global Book Productions, New York, 1992

"Packaging for the Mature Market"   Gerstman+Meyers Inc., *Qualitative Consumer Research Report*, New York, 1995

"The Store Strikes Back"   Paul Goldberger, *New York Times Magazine*, April 6, 1997

Guide to Packaging Law: A Primer for Packaging Professionals   Eric F. Greenberg, Institute of Packaging Professionals, Herndon, VA, 1996

"New Shades of Trademarking"   Eric F. Greenberg, *Packaging Digest*, June 1995

Marketing to and Through Kids   Selina S. Guber, Jon Berry, McGraw Hill, 1993

"The Relationship Between Trademark and Advertising"   Hall Dinkle Kent Friedman & Wood Inc., Memorandum to Clients, May 1995

Packaging Strategies: Meeting the Challenges of Changing Times   Arthur W. Harckham, Editor, Technomics Publishing Company, Inc., Lancaster, PA, 1989

The Total Package   Thomas Hine, Editor, Little, Brown and Company, 1995

"Corporate Identity Policies Need Overhauling in the '90s"   Murray J. Lubliner, *AMA Marketing News*, October 28, 1991

"The Age of Transnational Identity"   Murray J. Lubliner, *Public Relations Quarterly*, Winter 1991/92

"Are Outdated Identity Practices Undermining the Search for Quality?"   Murray J. Lubliner, *Graphic Design: USA*, June 1992

"Brand Name Selection: Critical Challenge for Global Marketers" — Murray J. Lubliner, *AMA Marketing News*, August 2, 1993

International Brand Packaging Awards I and II — Murray J. Lubliner, Editor, Rockport Publishers, 1993 and 1994

Global Corporate Identity — Murray J. Lubliner, Editor, Rockport Publishers, 1995

"Basic Answers to Those Typical Trade Mark Law Questions" — Markforce Associates, *Public Relations Bulletin*, London, 1997

Kids as Customers: A Handbook of Marketing to Children — James U. McNeal, Lexington Books, an Imprint of McMillan, Inc., 1992

"Packaging for the Elderly" — Herbert M. Meyers, Presentation, Pack Expo, New York, 1986

"The Ten Golden Rules of Private Label Packaging" — Herbert M. Meyers, Presentation, International Business Communications Conference, New Orleans, LA, 1993

"Recipe for Success" — Herbert M. Meyers, *Store Brands*, January/February 1993

"The Role of Packaging in Brand Line Extensions" — Herbert M. Meyers, *The Journal of Brand Management*, Vol. 1, Issue 6, Henry Stewart Publications, London, June 1994

"How to Design Elderly Friendly Packaging with Ageless Appeal" — Herbert M. Meyers, Presentation, Packaging Technology Conference, Las Vegas, NE, 1995

"Identifying Key Branding Strategies to Effectively Compete in Today's Marketplace" — Herbert M. Meyers, Presentation, The Marketing Institute Conference, Palm Beach, FL, 1995

"The Role of Design"   Herbert M. Meyers, Chapter XI,
                                         *Packaging Strategy: Winning the
                                         Consumer*, edited by Mona Doyle,
                                         Technomics Publishing Co., Inc.,
                                         Lancaster, PA, 1996

"America's Most Admired   Betsy Morris, *Fortune Magazine*,
Companies"   December 1996

"The Graying of Packaging"   Ben Myaris, Presentation,
                                         Packaging Strategy Conference,
                                         Atlanta, GA, 1993

"From Merchant to Marketer: Roundtable Report, Chicago, Private
Retailing's New Role"   Label Manufacturing Association,
                                         New York, November 1995

Marketing Warfare   Al Ries and Jack Trout, McGraw
                                         Hill, 1989

Fundamentals of Packaging Technology   Walter Soroka, Editor, Institute of
                                         Packaging Professionals, Herndon,
                                         VA, 1995

"Packaging for the Elderly"   Dr. Margaret Wylde, *ProMatura
                                         Perspectives*, a publication of
                                         ProMatura Group, a division of the
                                         Institute of Technology
                                         Development, 1995

"Eye Tracking Shows What Shoppers   Elliot Young, *Public Relations Bulletin*,
See and What They Miss"   Perception Research Services, Inc.,
                                         Ft. Lee, NJ, 1996

Index

About the Authors

Herbert M. Meyers has been a specialist in marketing, corporate identification, and package design for major corporations and design firms since 1955. Born in Germany, he graduated from Pratt Institute and gained substantial packaging experience while heading the design departments of major packaging manufacturers. As a vice president at design consultants Sandgren & Murtha, he supervised corporate and brand identity programs for major international corporations.

In 1970, Herb Meyers joined Richard Gerstman to become a founding partner of Gerstman+Meyers, a leading international consulting firm (now An Interbrand Consultancy) specializing in brand identity, package design, and corporate identification for more than 100 international corporate clients. He is a member and past president of the Package Design Council International, the Pan-European Design Association, the Design Management Institute, the Association of Professional Design Firms, and the American Marketing Association as well as a trustee on the boards of several educational organizations.

A recipient of numerous design awards, including several Clios and Package Design Council Gold Awards, he has been a frequent lecturer and writer on packaging and corporate identification subjects and a contributing author to several books, including

"Determining Communication Objectives for Package Design" for *Handbook of Package Design Research* (Walter Stern, editor) and "The Role of Design" for *Packaging Strategy: Winning the Consumer* (Mona Doyle, editor). Having retired from Gerstman+Meyers, he now serves as an independent consultant on brand identity and packaging.

Murray J. Lubliner has played a pivotal role in hundreds of packaging, brand, and corporate identity programs in the United States and overseas. He began his professional career more than twenty-five years ago at Lippincott & Margulies, where he helped develop many of the identity communications concepts that continue to be followed by designers and marketers. For almost twenty years, he was a founding partner in Lubliner/Saltz, Inc., an identity and design consultancy. Currently, he heads up Murray J. Lubliner Associates in New York, which specializes in marketing and communications consulting for consumer and business-to-business products, as well as business development and strategic planning for design consultancies.

Lubliner directs the International Brand Packaging Awards, the highly respected global competition that provides a forum for designers throughout the world to showcase their best work. He publishes and edits *Global Packaging & Brand Identity Report,* a monthly newsletter. He is a frequent speaker and writer on packaging and brand and corporate identity. His books include *Global Corporate Identity: The Cross-Border Marketing Challenge* and *International Brand Packaging Awards I and II.* He is an adjunct professor at New York University's Management Institute, where he teaches "Identity, Image, and the Bottom Line."

He received a B.A. in English and Psychology at New York University and did graduate work at Columbia University and the New School of Social Research.